SCIENCE SK[illegible]

Problems in GCSE Biology

J Bennetts
Senior Lecturer in Science Education
St. Martin's College
Lancaster

P Walters
Health Service
Carlisle

P Burdett
Head of Science
Thornes House High School
Wakefield

A Williams
formerly
Head of Science
Benfield School
Newcastle upon Tyne

Hodder & Stoughton
LONDON SYDNEY AUCKLAND TORONTO

Contents

Introduction iv
1 Information handling 1
2 Problem solving 17
3 Patterns 28
4 Evaluation 41
5 Synthesis 55
Talking about the answers 77
Information handling 77
Problem solving 79
Patterns 81
Evaluation 83
Synthesis 86

Note
The following symbols appear beside some questions in the text. See Introduction for details:
* Questions which require some knowledge of facts, concepts and principles.
A Questions for which a discussion of the answer is provided.

Contexts

The following table shows the context in which each question is set.

Context	Chapter 1	Chapter 2	Chapter 3	Chapter 4	Chapter 5
Nutrition: Animals	Q11 Q12 Q13	–	–	Q2 Q19	Q17
Nutrition: Plants	Q5 Q19	Q1 Q19	–	Q10	Q6 Q12 Q19
Transport: Animals	Q6	Q5 Q7 Q13	Q6 Q17 Q19	Q12	Q4
Transport: Plants	–	Q10	–	–	Q5 Q11
Respiration	Q17	Q8	–	Q19	Q3
Excretion	Q20	Q2 Q16	–	Q18	–
Cells & Growth	Q1 Q14 Q22	Q3	Q5 Q16	Q1	Q10
Response	Q2 Q7	Q9	Q12	Q7	Q7
Locomotion	–	–	Q15	–	–
Reproduction: Animals	Q18	Q6	Q4 Q14	Q15	Q18
Reproduction: Plants	Q3	–	–	–	Q14
Evolution	–	–	Q20	Q8 Q11	Q2 Q7 Q14 Q15 Q16
Microbes & Disease	Q15 Q23	Q12 Q14 Q15 Q17	Q9 Q18	Q6 Q17	–
Diversity of Living Organisms	Q9 Q10 Q25	Q18	Q1 Q2 Q3 Q7 Q11 Q13	Q5 Q14	Q1
Organisms & the Environment	Q4 Q8 Q16 Q21 Q24 Q26 Q27	Q4 Q11	Q8 Q10	Q3 Q4 Q9 Q13 Q16 Q19	Q3 Q8 Q9 Q12 Q13 Q17

Introduction: mainly for teachers

The **Science Skills** series is designed to cultivate the ability to handle information, solve problems and perform the other processes which are generic to science disciplines. National Criteria for GCSE Biology and other sciences require that no more than 25 per cent of the assessment be based on the recall of facts, theories, principles and concepts. Coursework is weighted at between 20 per cent and 30 per cent. The books in this series seek to develop and exercise all those attributes which make up the remaining 50 per cent of GCSE assessment.

We feel that it is important for students preparing for GCSE to experience biology in real contexts. Whenever possible, the questions in this book do this. National Criteria require that technological applications and social, economic and environmental issues pervade GCSE assessment in science subjects.

The rationale for the grouping of questions under our chapter headings is given in the introduction of **Science Skills: Problems in GCSE Science**. It is intended as a way of blocking together the objectives of National Criteria and relates to the epistemology of many syllabuses. However we freely admit that the placing of a question under a particular heading is somewhat approximate. Science does not operate in discrete packages. It is not easy to decide whether analysis of a set of data to find a rogue result is information handling or finding patterns.

The authors of the **Science Skills** series are committed to the aims of broad balanced science. **Problems in GCSE Science, Biology** (and soon **Chemistry** and **Physics**) attempt to give a coherence to the skills that are common to all the sciences. At present, most examinations for the Sciences Double Award contain some integrated questions. (This includes LEAG 'Combined' and MEG 'Combined', both of which have separate biology, chemistry and physics syllabus sections.) However the majority of questions in these examinations are identifiable as relating to 'Biology' or one of the other subjects. If all questions on blood had to involve some chemistry and physics, some of them would be very contrived. This book therefore looks at biology questions in the context of balanced science in which the interplay of common skills is fundamental.

With a few questions in this book the content naturally links to chemistry and physics and this development has been permitted. For example, if a student is to make sense of the biological effects of the discharge of hot water from power stations into rivers, it is reasonable that the causes of the power station heat discharge are considered. This indeed is the essence of GCSE Biology which takes account of technological, social, economic and environmental factors. Therefore the book will be of value to those following a GCSE Biology course irrespective of the balanced science issue.

The first four chapters begin with *multiple choice questions*, arranged in a gradation of difficulty. The questions which follow are also arranged in order of increasing difficulty. But it should be noted that there is a discontinuity. The multiple choice questions get progressively harder. The earlier non-multiple choice questions are then very easy.

Most of our questions are as 'content free' as possible. Those which require some knowledge of facts, concepts and principles are marked with an asterisk *.

A A number of questions are flagged with this letter. In the final chapter the answering of these questions is discussed.

In most cases, these are not simply mark schemes, but rather an analysis of what the question was designed to test and advice on how to approach it.

The four authors have all taught biology in comprehensive schools. Three are, or were, Heads of Science. One is a GCSE Chief Examiner and one is a hospital doctor.

1 Information handling

1 Cancer death rates in women

A comparison for 1950/1970 of the death rate for women in England from certain cancers (shown as a percentage increase or decrease)

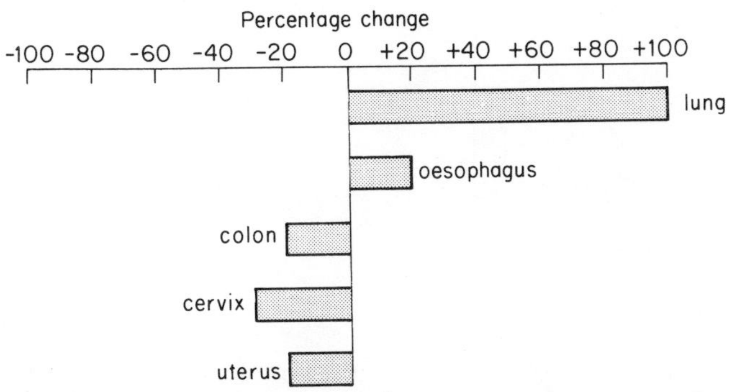

(a) Which type of cancer shows the greatest increase in the death rate for women from 1950 to 1970?
  A. Cervix.
  B. Colon.
  C. Lung.
  D. Uterus.

(b) Which type of cancer shows the greatest decrease in the death rate for women from 1950 to 1970?
  A. Cervix.
  B. Colon.
  C. Oesophagus.
  D. Uterus.

(c) Which of the following changes of women's behaviour may explain the trend shown in lung cancer and oesophageal cancer?
  A. Women smoking more.
  B. Women smoking less.
  C. Women drinking more alcohol.
  D. Women drinking less alcohol.

2 Diabetes

To find out if a person has diabetes or not, a blood sample is taken from them every 30 minutes for 2 hours. Glucose levels are found for each sample and plotted on a graph.

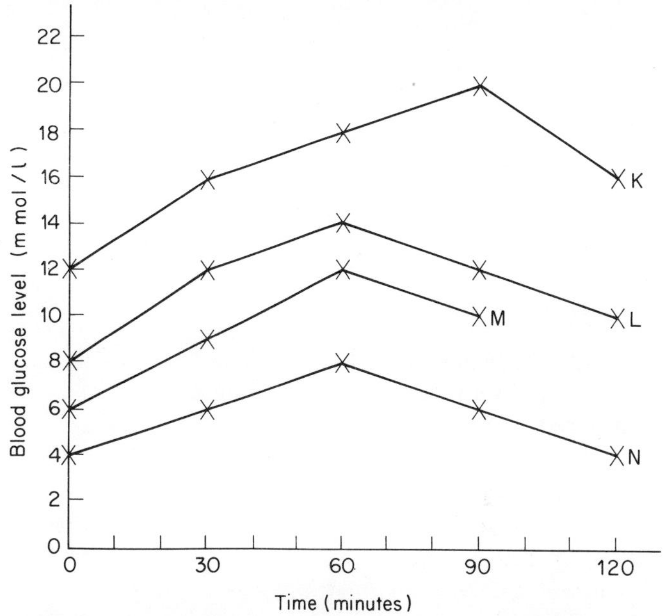

At time 0, the person was 'fasted' (starved of food overnight). A blood sample was taken and then the person was given 50 g of oral glucose for the test. If a person's blood glucose rises above 10 m mol/l, glucose starts to be detected in that person's urine. (The kidney is overloaded at this level.) Glucose in urine is very unusual.

Look at the results from 4 different people, K, L, M and N.

(a) The graph for a normal person is
  A. K.
  B. L.
  C. M.
  D. N.

(b) K's fasting blood glucose was
  A. 20 m mol/l.
  B. 16 m mol/l.
  C. 12 m mol/l.
  D. 4 m mol/l.

(c) After being given oral glucose, M would start to have glucose present in the urine after
  A. 15 min.
  B. 40 min.
  C. 1 hr.
  D. 1 hr 27 min.

A

3 Sycamore tree

This question is about the dispersal of seeds from a sycamore tree.

Sycamore 'helicopter' seed

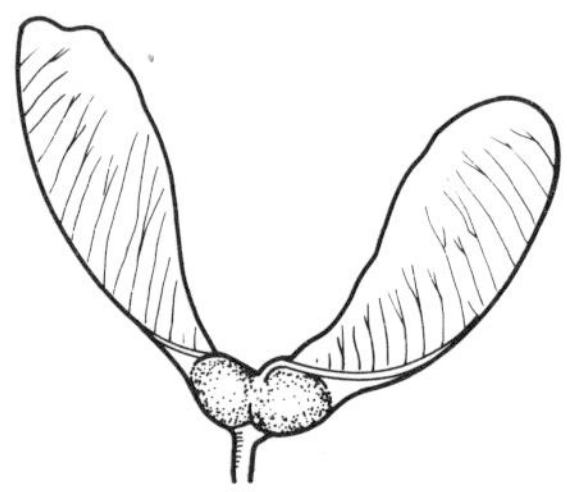

Eight directions are taken, each ten metres long, on eight points of the compass. These are numbered 1–8.

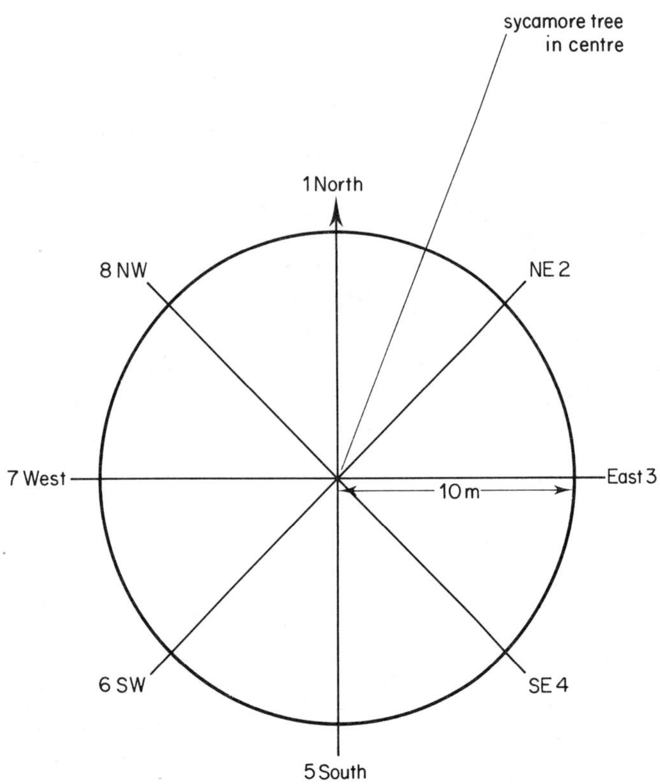

The number of seeds in a 50 cm × 50 cm quadrat is counted. The number of seeds found is converted into an 'abundance scale'.

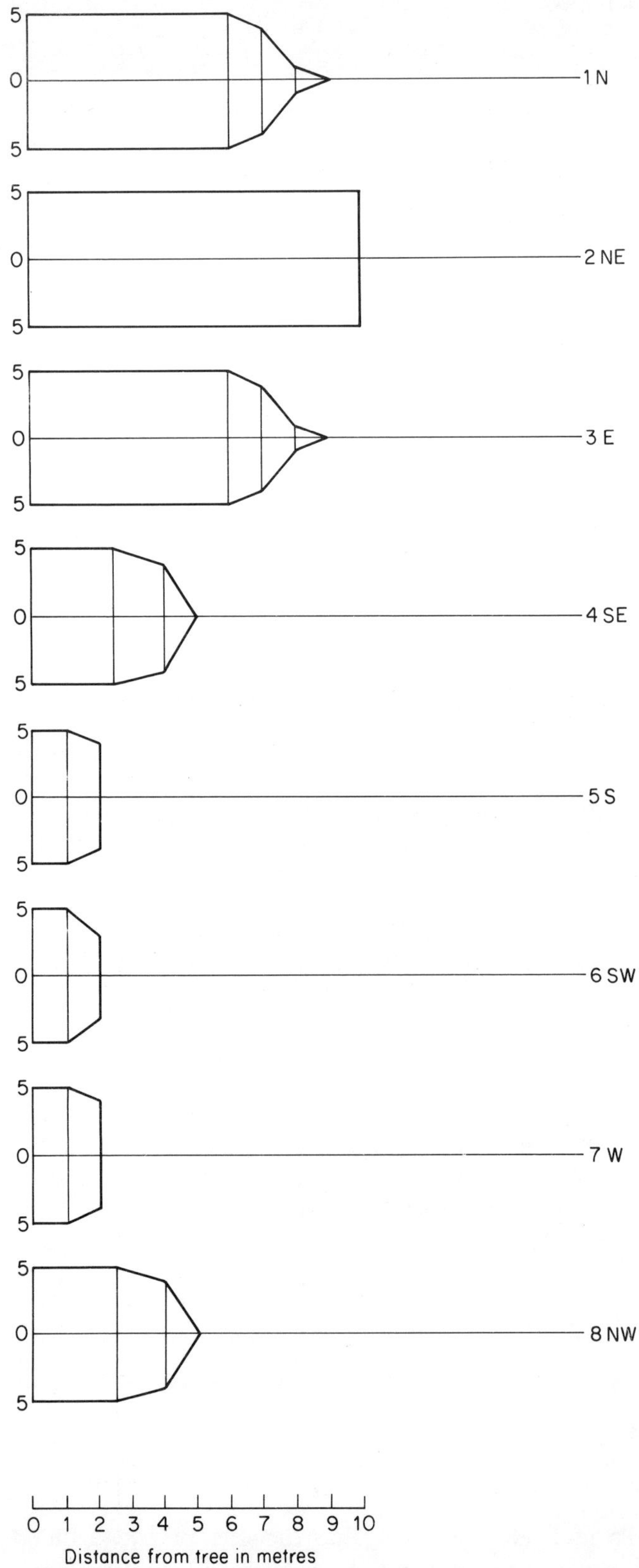

No. of seeds found	Abundance scale
21 or more	5
16 to 20	4
11 to 15	3
6 to 10	2
1 to 5	1

The kite diagrams on the previous page show how the abundance of seeds changes along each of the eight line directions.

(a) Which one of the following is most likely to alter distribution of sycamore seeds?
  A. Ice.
  B. Snow.
  C. Rain.
  D. Wind.

(b) The data suggest that when seeds are falling the wind is mainly from the
  A. SW. (south west)
  B. NE. (north east)
  C. W. (west)
  D. E. (east)

(c) Looking at the kite diagrams, a student reasons that the average length of branches on this sycamore tree is 2 m. Which explanation backs this up best?
  A. The abundance scale on each diagram is the same at 1 m.
  B. No matter which direction is looked at there are always seeds in a 2 m radius around the tree.
  C. The branches must be 2 m in length for seeds to drift up to 10 m away.
  D. No seeds are found further than 2 m from the tree.

(d) Which of the following sketches best represents the seed cover around the tree?

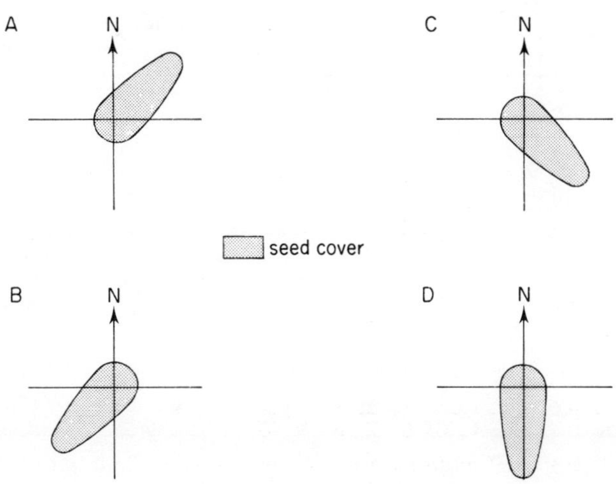

*4 Sewage outlet

The four graphs show changes in the chemicals and living things in a river above and below a sewage outlet.

The effects on a river below a sewage outfall

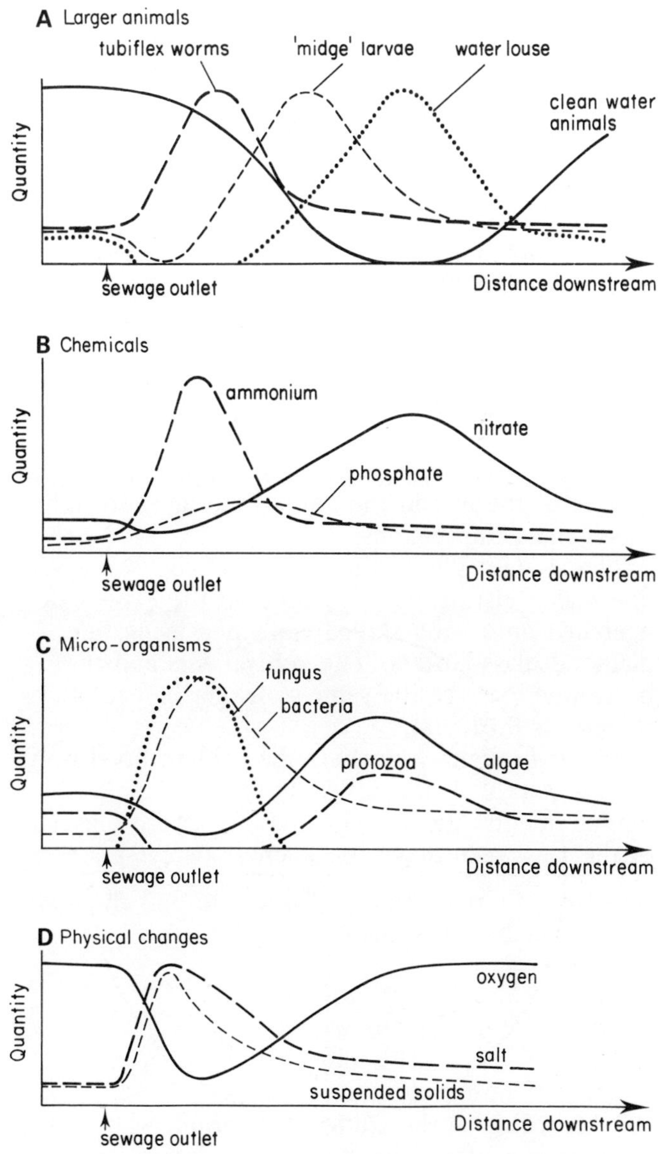

Read through the statements below and write down the letter(s) of the graph(s) that tell you this information.

(a) Sewage contains salt and suspended solids.
(b) Oxygen is used up by bacteria in the water.
(c) Bacteria produce ammonium and nitrate ions.

(d) Algae thrive and grow on the nitrate ions.
(e) These algae are eaten by protozoa which increase.
(f) Some animals live only in clean water with high oxygen levels.

5 Apple leaf

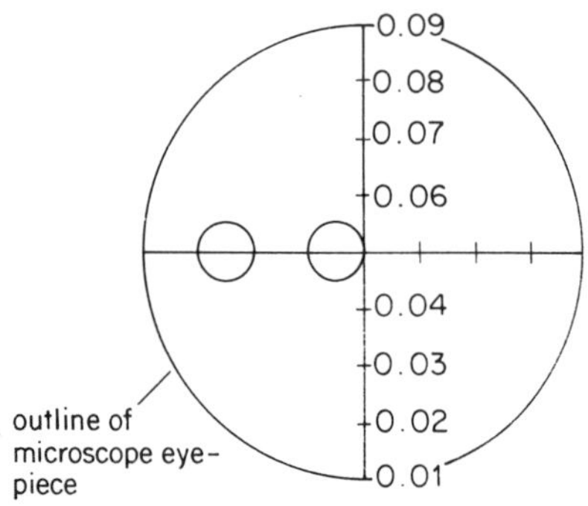

This is a scale diagram. A student drew it after making a 'Sellotape' copy of the undersurface of an apple leaf. When 'Sellotape' is stuck to a leaf and then gently removed, it leaves an impression of the surface of the leaf in the glue of the tape, including the stomata impressions.

The student placed the 'Sellotape' copy on a very fine scaled glass slide. The scale on the glass slide is etched onto it by skilled craftsmen using fine diamond glass cutters. The vertical axis and the horizontal axis are the same scale. Each space represents 0.01 mm.

The two circles are where the student lined up two stomata with the scale. By using careful measurements, the student drew to scale a diagram of the view seen down the microscope eye-piece.

(a) Use your ruler on the diagram to find the scale chosen by the student.
The scale is
A. 0.5 mm ≡ 0.01 mm.
B. 0.5 cm ≡ 0.1 mm.
C. 0.5 cm ≡ 0.01 mm.
D. 0.5 mm ≡ 1 mm.

(b) According to the student's diagram, what is the diameter of a stoma of an apple leaf?
A. 0.005 mm.
B. 0.01 mm.
C. 0.02 mm.
D. 0.1 mm.

(c) Assume that the distance between the edges of the stomata shown above is the average distance between all the stomata. How many stomata impressions would there be if the piece of tape used measured 1 cm × 5 cm?
A. 1000.
B. 5000.
C. 250 000.
D. 1 250 000.

*6 Blood counts

4 to 11	White Blood cells (WBC) × 10^9/l
4.5 to 6.5	Red Blood cells (RBC) × 10^{12}/l
150 to 350	Platelets × 10^9/l
12 to 18	Haemoglobin (Hb) g/dl
76 to 100	Mean or average cell volume (MCV) fl

Print out no. 1

This is print out no. 1 from an automatic blood cell counter. The normal range of results for a healthy person are printed in the left hand column.

There are between 4×10^9 and 11×10^9 (4–11 thousand million) white blood cells in a litre (l) of blood.
There are between 4.5×10^{12} and 6.5×10^{12} (4.5–6.5 million million) red blood cells in a litre of blood.
There are between 150×10^9 and 350×10^9 (150–350 thousand million) platelets in a litre of blood.
There are between 12 and 18 grams of haemoglobin in a decilitre (dl) of blood (100 cm^3).
The average volume of red blood cells is between 76 and 100 femtolitre (fl; 10^{-15} litres)

Doctors use these results to help diagnose disease. Look at the print outs below of blood sample results from four different patients.

Mr V		*Mr W*		*Mr X*		*Mr Y*	
6.5	WBC × 10^9/l	1.0	WBC × 10^9/l	4.1	WBC × 10^9/l	200	WBC × 10^9/l
3.3	RBC × 10^{12}/l	5.2	RBC × 10^{12}/l	3.0	RBC × 10^{12}/l	5.9	RBC × 10^{12}/l
200	Platelets × 10^9/l	210	Platelets × 10^9/l	70	Platelets × 10^9/l	10	Platelets × 10^9/l
6.2	Hb g/dl	13.1	Hb g/dl	5.0	Hb g/dl	12.0	Hb g/dl
80	MCV fl	90	MCV fl	125	MCV fl	82	MCV fl
Print out no. 2		Print out no. 3		Print out no. 4		Print out no. 5	

(a) Which patient is most likely to complain of tiredness and becoming breathless when climbing upstairs but has cells which look normal size on a blood film?
A. Mr V.
B. Mr W.
C. Mr X.
D. Mr Y.

(b) Which patient is most likely to complain of spontaneous bleeding and bruising?
A. Mr V.
B. Mr W.
C. Mr X.
D. Mr Y.

(c) Which patient complains of many sore throats which take a long time to get better, and that he has had two chest infection attacks in one month?
A. Mr V.
B. Mr W.
C. Mr X.
D. Mr Y.

(d) Leukaemia means 'white blood'. This is because the blood film shows many white blood cells. Which patient may have leukaemia?
A. Mr V.
B. Mr W.
C. Mr X.
D. Mr Y.

7

200 people were questioned about the quality of their eyesight. The pie chart below shows the results of the survey.

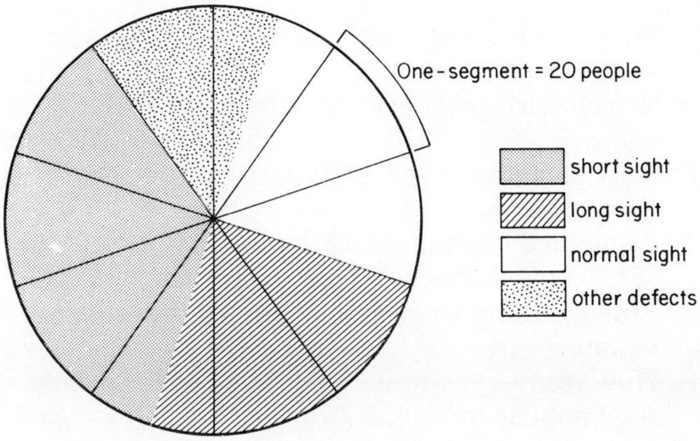

(a) How many people had
(i) short sight? *(1)*
(ii) normal vision? *(1)*

(b) Are these two groups together more than the rest? *(1)*

(c) What was the total number of people asked? *(1)*

(d) What percentage had normal vision? *(2)*

8

Sukhvinder throws a 0.25 m^2 quadrat on part of a playing field. The diagram below shows what she found.

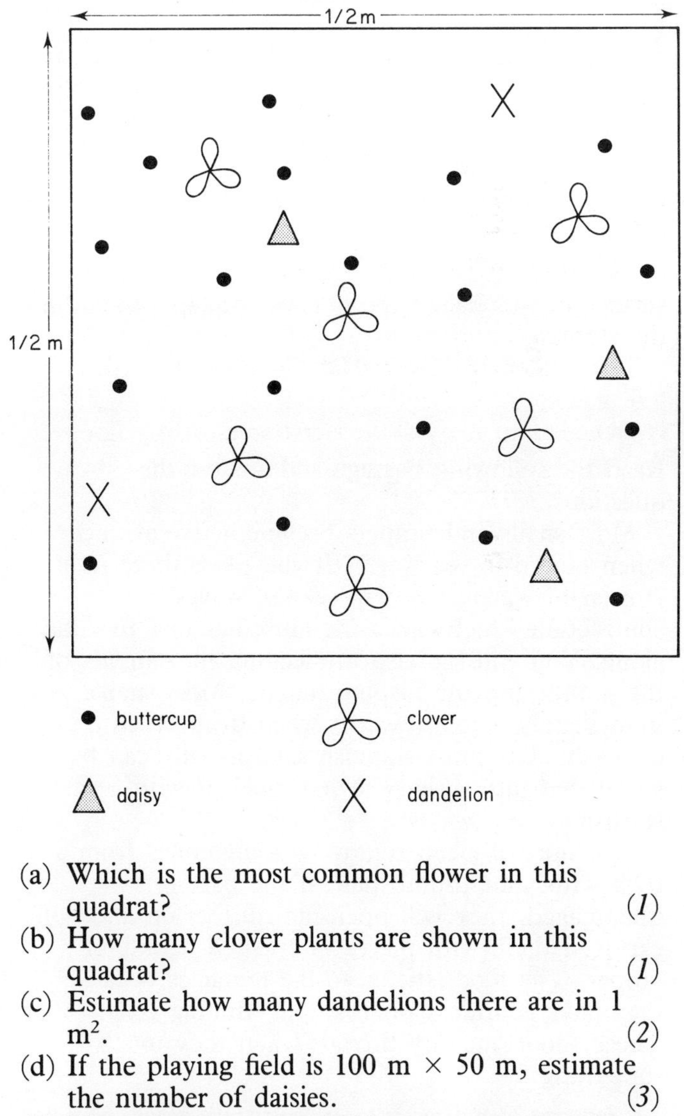

(a) Which is the most common flower in this quadrat? *(1)*

(b) How many clover plants are shown in this quadrat? *(1)*

(c) Estimate how many dandelions there are in 1 m^2. *(2)*

(d) If the playing field is 100 m × 50 m, estimate the number of daisies. *(3)*

(e) These figures are probably not very accurate. What should Sukhvinder do to make her results more reliable? *(1)*

9

This table shows the months when certain plants are in flower.

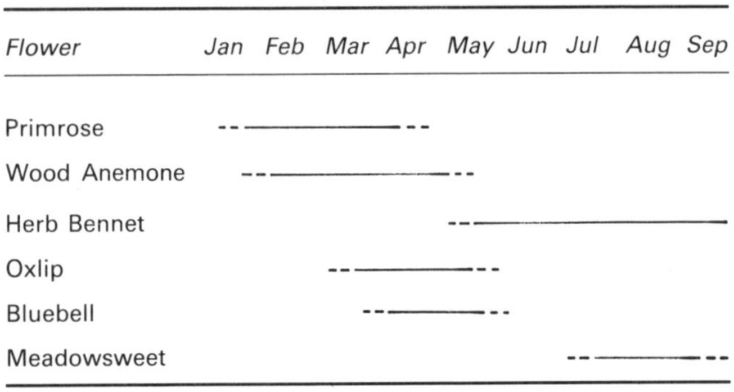

(a) Put the flowers in order of flowering. Put the one that flowers the earliest, first. *(3)*
(b) Which one flowers longest? *(1)*
(c) Which season, spring or summer, has the most plants flowering? Give a reason. *(2)*

*10 Seashore life

Read the following passage and answer the questions.

Most snails and limpets become active at once when placed in sea water. If you place them in a jam jar of water, you will see the waves of contraction which cause the muscular foot to glide along. You will see also the feeding movements of the mouth and the rasping tongue. Most snails immediately crawl upwards when dropped in a jar of water. The most sluggish snail usually can be stimulated into activity if you touch it with a small starfish.

You normally can remove sea anemones from a rock with your thumb nail. If the base is undamaged, they will open out quite soon and you can feed them with pieces of crushed winkle or limpet. The food 'sticks' to the tentacles because they have countless microscopic stinging cells. These shoot out tiny threads when they touch something.

(a) What word is used to describe the limpet's tongue? *(1)*
(b) Limpets are herbivores. What do you think they might feed on? *(1)*
(c) What do most snails do when placed in a jar of water? *(1)*
(d) Suggest two reasons why they might react like this. *(2)*
(e) What is a common predator of snails? *(1)*
(f) Suggest how sea anemones stay attached to rocks. *(1)*
(g) How do sea anemones catch their prey? *(1)*

*11 Eggs as food

This is the label from a pack of hen eggs.

Contents	*per 100 g*	*per egg*
Energy	145 kcal 610 kJ	80 kcal 340 kJ
Protein	12.3 g	6.8 g
Carbohydrate	0.4 g	0.2 g
Total fat	10.9 g	6.1 g
Polyunsaturates	1.8 g	1.0 g
Saturates	3.0 g	1.7 g
% of human's daily requirement		
Vitamin D	70%	40%
Vitamin B_2	30%	15%
Vitamin B_{12}	85%	45%
Vitamin A	15%	8%
Iodine	35%	20%

Use this information to answer the questions. (Most of the answers are one word or short sentences.)

(a) How much energy will one egg give you, if it is a boiled egg? *(1)*
(b) Which part of an egg contains most of the energy? *(2)*
(c) Why should the energy supply for a chick be in this form? *(1)*
(d) Name one vitamin which is missing. *(1)*
(e) What must a chick be able to do that humans cannot in order to get a supply of this missing vitamin? *(1)*
(f) How many eggs would supply a human's daily requirement for iodine? *(1)*

(g) Would it be a good idea to eat this number of eggs every day to get the right amount of iodine? (Think about the other contents of the egg.) *(2)*
(h) How much carbohydrate is there in one egg? *(1)*
(i) Why is there so much protein inside an egg? *(1)*
(j) What would the calcium and vitamin D be needed for? *(3)*
(k) A chick takes three weeks to develop. There is not much energy in the egg for this length of time. Write down two ways in which the chick might have a reduced energy demand. *(2)*
(l) Where is the water supply stored in an egg? (This is a 'looking up' answer.) *(1)*
(m) Find out what vitamin B_{12} is used for. Why is it so important to a developing chick? *(1)*
(n) Will two eggs give a whole day's supply of vitamin B_{12}? *(1)*
(o) Do eggs contain more energy at 340 kilojoules each than at 80 kilocalories each? *(1)*

12 Caterpillar

Some pupils were keen to investigate how quickly caterpillars could eat different types of leaf.
Diagram A shows a soya bean leaf before being eaten.
Diagram B shows the soya bean leaf after 2 minutes.

1 square ≡ 2 cm²

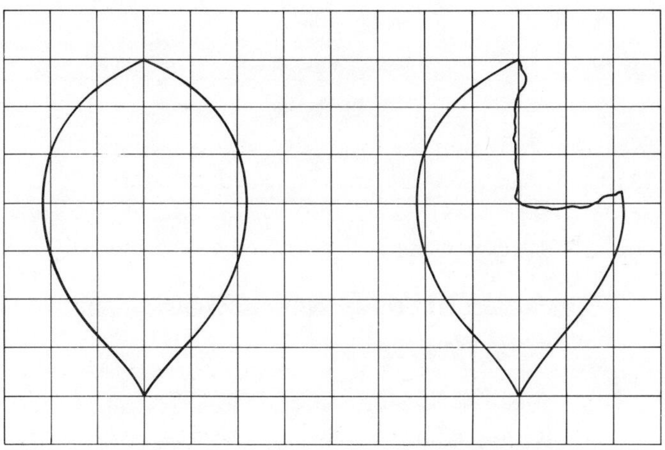

Diagram A Diagram B

(a) What is the area of the leaf before being eaten? (To the nearest whole cm².) *(1)*
(b) What is the area left after 2 minutes? *(1)*
(c) How many cm² have been eaten in 2 minutes? *(1)*
(d) What fraction or percentage of the leaf has been eaten? *(1)*
(e) How long would it take to eat the whole leaf? *(1)*
(f) How many leaves could the caterpillar eat in 32 minutes? (2)
(g) The actual number of leaves eaten in 32 minutes is less than that calculated in (f). Suggest two reasons why. (2)
(h) If the pupils wanted to compare the rate at which caterpillars eat different types of leaf, what two things must be kept the same? (2)

13 Chewing bread

Jane was told by her biology teacher that she should chew her food well before swallowing it. She was told that this was because her food was better digested and caused 'fewer problems with her gut'.

Jane did not believe the teacher and so she decided to do a simple experiment to find out for herself. She chose bread. She cut several equal-sized cubes of bread. Jane chewed a cube for 5 seconds and then spat it into a container. She then tested this chewed bread for starch.
She repeated this, chewing the bread longer each time. She also tested the bread without chewing it and found that it was very high in starch.

The results were:

Time in seconds	0	5	10	15	20	25	30	35	40	45
% of starch left in the chewed bread	100	80	60	40	30	22	16	14	14	14

(a) Plot a line graph of these results. *(4)*
(b) What % of starch remains if the bread is chewed for 8 seconds? *(1)*
(c) How long would she have had to chew for 18 per cent to remain? *(1)*
(d) What chemical could be used to test for starch? *(1)*
(e) Was the teacher right? Explain your answer. (2)
(f) Saliva contains a chemical that breaks down only starch. Do you think that the results would have been the same if she had chewed meat and tested for protein? Explain your answer. (2)

(g) How does this affect your answer to part (e)? i.e. Is the teacher as right as you first thought? Explain your answer. *(3)*

14 Hair

Hair, finger and toe nails, some horns (e.g. rhino) and wool are all made of a substance called keratin. Some layers of skin contain keratin.

Facts about keratin/hair

Solubility in water:	low
Tensile strength (force needed to break it):	high
Effect of hot water:	softens it
Rate of growth on head:	0.35 mm/day

Hair grows out of a special group of cells in folds in the skin called hair follicles. A straight follicle produces straight hair and a curved follicle produces wavy hair.

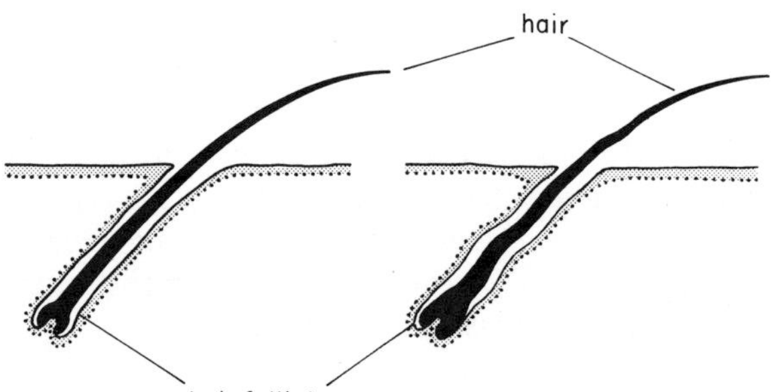

Different parts of the body have different types and amounts of hair. Some peoples, such as the Ainu tribe in Japan, have more body hair than most. Some people have fine hair like Scandinavians, whereas African hair is much thicker. Some religions do not permit hair to be cut, like Sikhs.

(a) Suggest explanations for the following (based on the structure of skin or hair).
   - (i) Armpit hairs are much thicker than eyelashes. *(1)*
   - (ii) Some parts of the body have more hair than others. *(1)*
   - (iii) Skin on fingertips becomes wrinkled after doing a lot of washing by hand or after a long bath or shower. *(1)*
   - (iv) It is easier to cut your nails after you have been in a bath. *(1)*
   - (v) Knife handles and ornaments can be made out of horn. What could you do to the horn to make it easier to cut and shape? *(1)*
   - (vi) Some hair salons wrap a hot, damp towel around a man's face before shaving him. *(1)*
   - (vii) Men have both thicker hair and more hair on their face than women. *(2)*
   - (viii) Shaving foam consists of bubbles that trap water. How does this make shaving beards easier? *(1)*

(b) How many days would it take a person to grow his or her scalp hair by 3.5 cm? *(1)*

15 Decomposing

Study the diagram below which compares the amount of leaves decomposed in a soil at different temperatures.

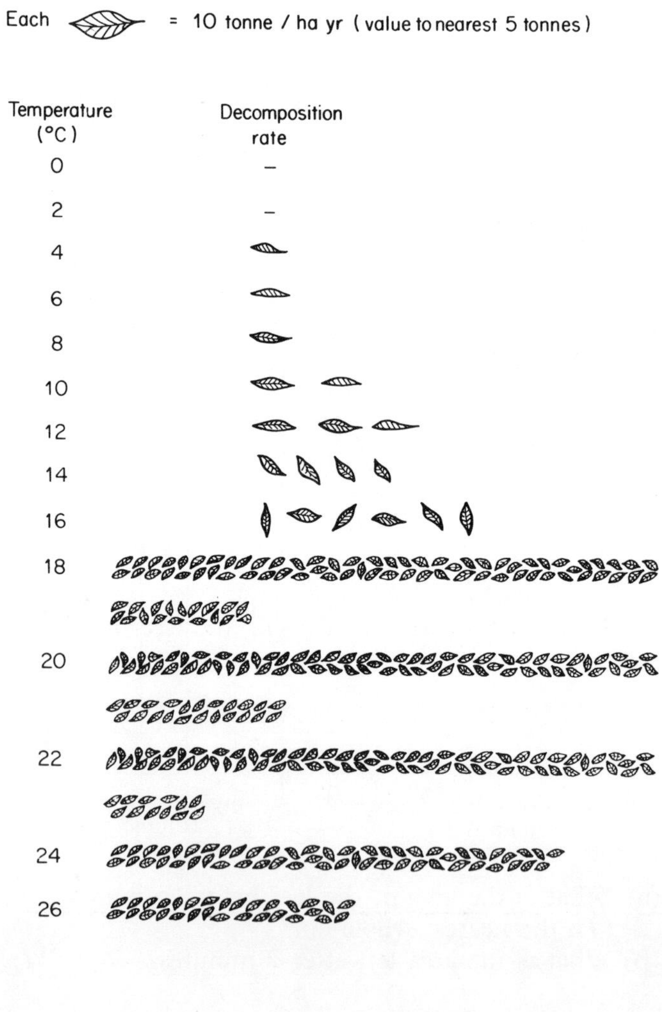

(a) Plot a line graph with decomposition rate on the vertical (upright) axis and temperature along the bottom. Draw the 'best average line' through the points. *(5)*

(b) Write in your own words a description of the change in decomposition rate with temperature. *(3)*

(c) How much is decomposed at
  (i) 8°C,
  (ii) 12°C,
  (iii) 18°C? *(3)*

(d) Suggest an explanation for the breakdown rate at 0°C. *(1)*

(e) Predict how much will be decomposed at 28°C. *(1)*

(f) What temperature might sterilise soil to prevent any decomposition of leaves? *(1)*

16 Biomass

The diagram below shows a pyramid of biomass for a foodchain in a plantation in the Caribbean.
Scale: 1 mm = 100 kg.

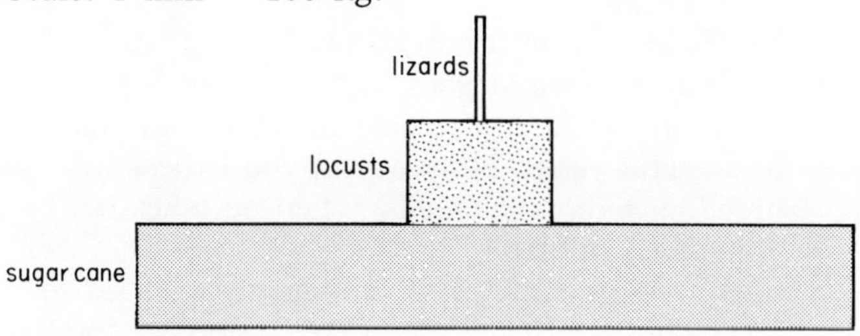

(a) What is the biomass of
  (i) locusts,
  (ii) sugar cane,
  (iii) lizards? *(3)*

(b) What
  (i) fraction
  (ii) percentage
  of the mass of sugar cane ends up as lizard biomass? *(3)*

(c) Draw your own pyramid using the data below. Choose your own scale.

Cicada grasshoppers	2500 kg
Olive trees	10 000 kg
Banded snakes	1000 kg
Eagles	50 kg

(6)

17 Breathing

Some pupils wanted to investigate the effect of different gases on the breathing of an insect. They decided to use a locust apparatus like that shown below.

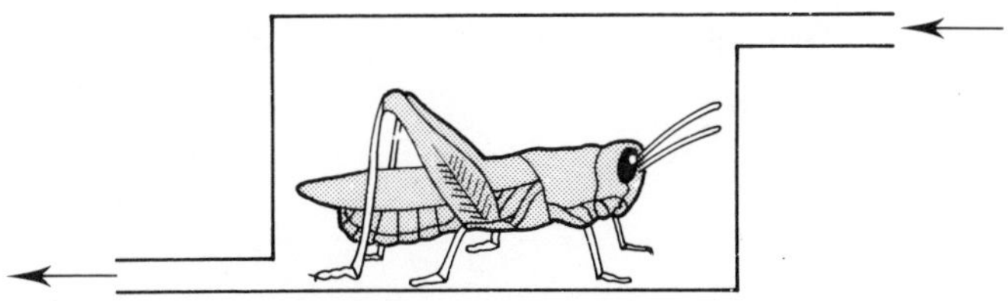

Various gases were passed gently over the locust and the breathing movements of the abdomen were counted for 15 seconds.

(a) Copy and fill in the table below.

Gases	*No. of breaths in 15 seconds*	*No. of breaths in one minute*
Air	5	
Exhaled air	9	
Pure oxygen	2	
Pure carbon dioxide	10	
Dentist's anaesthetic	2	

(5)

(b) Plot a simple bar graph with breaths per minute on the vertical axis. (5)

(c) Explain why exhaled air caused such a high rate. (2)

(d) Explain why pure oxygen caused such a low rate. (2)

(e) Explain why dentist's anaesthetic should have caused a similar breathing rate to that with oxygen. (2)

(f) Why was it necessary to include the normal air result? (2)

18 Cats

When a female cat (a queen) comes on heat she starts to make loud, long and high pitched cries. This 'calling' brings one or more tom cats (males).

The female rolls on the ground and moves in an unusual way, creeping along the ground with her tail in the air. This tells the tom cat that she wants to copulate. The tom mounts the female, sometimes holding the fur at the back of her neck with his teeth and intercourse takes place.

The tom sprays all trees and fences in the female's garden with urine. This contains a substance with a musky smell. This warns other toms to keep away. The father may fight other tom cats which come near the territory.
The female ovulates immediately after intercourse. The gestation period is 9 weeks. In England female cats usually produce kittens in early April and again in July. Young kittens suckle from the mother's teats. After about six weeks the mother may catch mice for the kittens to kill and eat.

When kittens are two months old they are able to fend for themselves. Owners usually find new homes for them at this stage. If a kitten comes back a few months later, the mother may fight it, like any other cat straying on the mother's territory.

(a) At what time of year do cats come on heat in England? (2)
(b) Females normally turn and scratch tom cats which sniff round their tails at the wrong time. How do queens tell toms that they are ready for sexual intercourse? (3)
(c) Owners of female cats may not always see them copulating. But it is important for them to know when to expect kittens. How can the owners tell when copulation has taken place? (1)
(d) Females might copulate with two different tom cats during the same afternoon. Toms try to stop it. How might they do this? (2)
(e) Suggest reasons why the mother chases away kittens which left six months earlier. (2)
(f) When does the second set of kittens become independent of the mother? (1)
(g) If kittens are born in England during December, they may not survive as well as kittens born during July. Explain this. (1)
(h) What seasonal changes might control the timing of a female coming on heat? (2)
(i) Cats do not usually come on heat for two weeks after having the first set of kittens in the year. Suggest reasons for this. (2)
(j) In New Zealand the shortest days are in June. The coldest time of year is July and August. When would you expect cats to come on heat in New Zealand? (2)
(k) Trinidad is near the Equator. It becomes light at about 6 am and goes dark at about 6 pm. Winston has kept a female cat in Birmingham for several years. How would you expect its breeding habits to change when he takes it home to Trinidad? (1)
(l) Cats are sometimes kept on ships to get rid of rats. They are usually neutered. Suggest reasons for this. (2)

*19 Sugars

This is a comprehension question. Read it carefully and then answer the questions.

The name sugar is used for several kinds of sugar. The sugar that we use for cooking and at table has a scientific name, *sucrose*. There are other sugars too. Honey is very sweet because it contains fruit sugar or *fructose*. Brewers use malted grains because brewers' yeast can feed on the *maltose* in malt. *Lactose* is sugar found in milk. Most living cells rely on *glucose* sugar for energy supplies.

All sugars are made by plants, except for lactose. This is made by the mammary glands of female mammals. The female mammals get their sugar from food. Their food has come from plants at the start of a food chain. Plants make sugars when they carry out photosynthesis. Photosynthesis is a set of chemical reactions that happen inside green leaf cells. The green leaf cells absorb some energy from sunlight and use it to make sugar from carbon dioxide (CO_2) and water (H_2O). Here is a diagram to show what a leaf cell needs.

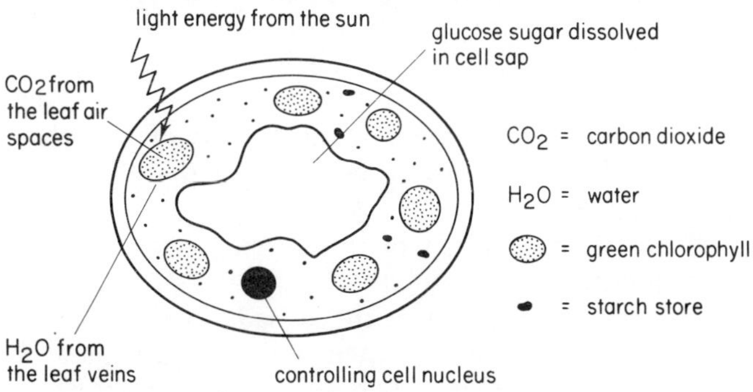

You will see that the glucose sugar made is dissolved in the cell sap. In some plants, for example the green stripey *Chlorophytum* plant, this sugar is always stored in the cell sap. But usually it is changed into starch very quickly. Starch is made of giant molecules that are built up in plant cells from strings of glucose molecules. Starch molecules are long and coiled. They are for storing energy-containing glucose molecules. This is why we find starch and not sugar when we test a green leaf to see whether it has carried out photosynthesis. At night time, when there can be no photosynthesis because it is dark, the starch is changed back into smaller glucose molecules. This is moved from the leaves to special storage places like the centres of stems and roots. Sometimes it is stored as sucrose.

(a) What is the scientific name for the sugar that we use at home? *(1)*
(b) Which sugar is very sweet? Which part of a plant will probably contain this kind of sugar? Name an example and suggest a reason why this sweetness is necessary for the plant. *(4)*
(c) Which sugar is found in a food that comes from animals? *(1)*
(d) Write a food chain that starts with the sugar in sugar cane and ends with a newly born panda. *(2)*
(e) Leaf cells are like small chemical factories. They make sugar from raw materials. Where does the leaf's energy supply come from? *(1)*
(f) Use reference books to find out whether a fungus can make its own sugar. *(3)*
(g) Explain the sentence, 'At night time, when there can be no photosynthesis because it is dark . . .'. How does darkness affect photosynthesis? *(1)*
(h) Draw a plant outline and work out the transport patterns for sugar which is being moved from the leaves to storage places. *(2)*
(i) Greenfly have pointed mouthparts like needles. When they suck plant juices they take too much sugar and excrete it as honeydew. They plunge their mouthparts into the phloem tissue of stems. What does this tell you about sugar transport through stems? *(1)*

*20 Kidney

This table shows data measured for one kidney over 24 hours.

	Litres per day
Total blood entering kidney	800.0
Blood plasma entering kidney	500.0
Fluid filtered out	90.0
Urine leaving kidney	0.8

(a) What makes the difference between blood plasma and total blood volume? *(1)*
(b) What percentage of the total volume of blood entering the kidney leaves as urine? *(1)*
(c) Estimate the volume of blood leaving both kidneys in a day. *(1)*
(d) 90 litres of fluid are filtered per day yet only 0.8 litre of urine is formed. What happens to the rest? *(2)*
(e) Which values would change most for the following situations? Explain your answers.
  (i) A very hot day. *(2)*
  (ii) Moderate exercise on a cold day. *(2)*
  (iii) Eating a salty meal. *(2)*

21 Conserving the Norfolk Broads

The Norfolk Broads in Eastern England are large, fairly shallow, lakes. They have a rich variety of wildlife, but this is being threatened by people's activities. The Broads are popular for pleasure-boating and fishing. Fishing alters the balance of fish populations both as a balance between different species and as a balance between different ages of fish. The water is becoming polluted and boatwashes are damaging the banks. This diagram shows how different human activities are affecting the natural state of these water habitats. Several plant and animal species are mentioned and there are brief notes on them below.

Coypu: large rodents, like giant voles, introduced from South America into fur farms to make inexpensive Nutria coats. Now escaped and feeding on the reeds.
Prymnesium parvum: a poisonous alga that kills fish.
Phytoplankton: floating plant life, (consisting of very small algae) which gives the water an appearance of pea soup and cuts out light from plants below.
Botulism: food poisoning, caused by bacteria that flourish in sewage and bird life.
Peucidanum palustre: food plant of the Swallowtail butterfly larva, (*palustre* = marsh dwelling).

(a) Which four human activities have adversely affected the Broads? (*4*)
(b) Trace through the diagram to find out the effects of intensified agriculture. (*6*)
Source: *Biologist* (1987, Vol 34, No. (1)).
(c) The foodplant of the Swallowtail butterfly larva will grow only in specialized habitats. Show how human activity can cause the near extinction of Swallowtails in the fens around the Broads. (*3*)
(d) How are fish stocks affected by the increased population? (*4*)
(e) Point out three effects in the Broads of growth in the leisure industry. How might these be self-defeating? (*3*)
(f) What are the effects of introducing coypu by accident? (*3*)
(g) Use this description of the Broads in the years 1900–1925 to contrast them with the Broads at present. How many changes can you identify? (*4*)

'The first freshwater biological station was founded on the Broads in 1902. Work was done on Crustaceans, water beetles and fish. Local industries

Interrelationships of the Broadland problems

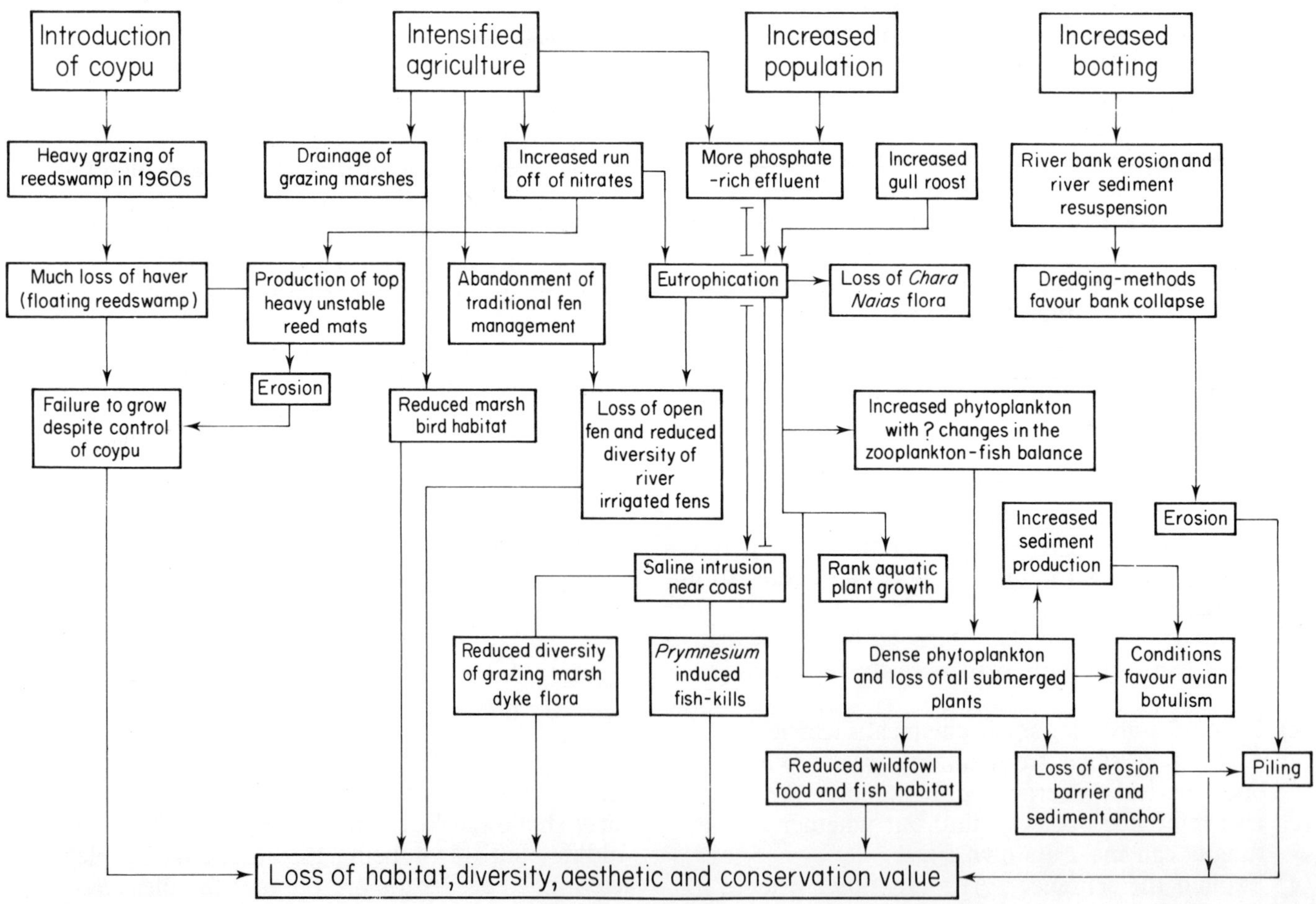

included eel fishing, reed-cutting, sailing and catering for serious naturalists. Records show that they studied the rivers and wet fens. They found Swallowtail butterflies, bitterns and marsh harriers and a variety of local birds. The waterways were so clear that stoneworts could grow in them.'

*22 Mutations

Genetic 'instructions' inherited by children from their parents are in the form of *genes*. Most genes work in pairs (one from the father and the other from the mother). Not all genes are identical. In any population one gene may have several forms, called *alleles*. The A,B,O blood grouping system has three alleles called A,B and O, but there can be only two alleles in every individual. This gives a variety of blood groups in the human population.

(a) Write down all the possible allele pair combinations that people could have. (*3*)

(b) Alleles are the result of changes to or mutation in the genes. They happen naturally but can be increased by exposure to radiation. Many brave firemen were exposed to radiation at Chernobyl in April '86. Scientists are now calculating the number of mutations per person that may have occurred. They looked at the effects of α radiation and β radiation.

Length of time exposed to radiation (hours)	*Estimated number of mutations per person*	
	α	*β*
0	5	5
½	5	10
1	5	20
1½	5	40
2	5	80
2½	5	85
3	5	90
3½	10	90
4	15	90
4½	20	90
5	25	90
5½	30	90
6	35	90

Plot a line graph with two lines, one for α and the other for β. Put the length of time on the horizontal (bottom) axis. This axis must extend to a time of twelve hours.

(i) Why are there still mutations when the exposure time is 0? (*1*)

(ii) Briefly summarize what each line on your graph shows. (*4*)

(iii) Which radiation is the most dangerous and why? (*1*)

(iv) What can you say from the lines about the skin penetrating power of α and β radiation? (*1*)

(v) From the graph, estimate the time of exposure for which the number of α mutations is equal to the number of β mutations. (2)

(vi) The actual number of mutations per person is turning out to be much less than the graph shows. Can you suggest two reasons why? (2)

*23 Microbe leachers

Read the passage carefully before you answer the questions.

In the USA almost 10 per cent of all copper metal is obtained by using microbes, which enable copper to be removed cheaply from low grade ores. The bacteria used come from the naturally occurring *Thiobacillus* group. They are encouraged to grow by providing good aeration and spraying acidified water over the ore. The bacteria convert the iron found in copper ores from ferrous (iron II) to the ferric (iron III) form. They obtain energy when they do this. Iron III is an oxidising agent and oxidises copper sulphide (pyrite ore) to form sulphuric acid. This maintains the acidity of the leaching solution. The iron III also oxidises insoluble pyrite ores to produce copper sulphate. This is soluble and runs out of the bottom of the dump where it can be pumped to a factory. Copper is then recovered from the solution by electrolysis.

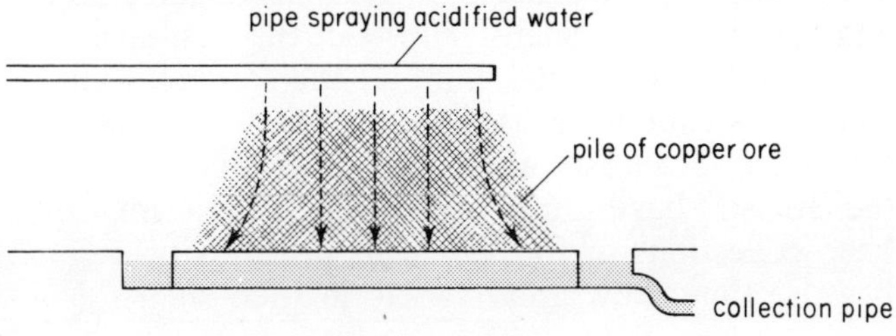

(a) Is *Thiobacillus* aerobic or anaerobic? (*1*)
(b) Does it prefer surroundings at high or low pH? (*1*)
(c) What is the chemical name for iron pyrite? (*1*)
(d) Why might this method be more efficient in a warm area than a cold one? (*1*)
(e) What other metal might be found in the 'liquor' produced at the bottom of the dump? (*1*)
(f) What is the order of the following reactions in the passage?

A. ironIII + copper + oxygen + water → sulphide
sulphuric acid + ironII + soluble copper
B. ironII → ironIII
C. ironIII + copper sulphide + oxygen → copper sulphate + ironII

24 Saving a snail

This is a comprehension question. Read it carefully and then answer the questions.

Moorea is a tropical island in the Pacific Ocean. Humans have created a new problem with the wildlife on the island by altering the balance of nature. For a million years, the *Partula* snail has lived in the thick green forests on the island. Now it is threatened. People living on the island like to eat snails, but not the *Partula* snail. They prefer the Giant African Land snail. So they started snail farms to grow the Giant African Land snail. Some of the Giant Land snails escaped. They became pests because they climbed up walls and ate crops. Scientists found that a snail that lives in Florida liked to eat the Giant Land snails, so some Florida snails were imported to Moorea. When the Florida snails were set free in the island forests and fields, they attacked the Giant Land snails, but the Florida snails soon discovered that the *Partula* snails, who had been living on the island for so long, were much more tasty so they began to eat them too.

Now there are several problems: too many Giant Land snails, far too many Florida snails and fewer and fewer *Partula* snails. How can the pest snails be controlled without killing off the *Partula* snails? Chemicals cannot be used because they will kill all the snails. Scientists who study inheritance (geneticists) have found an answer. They want to keep collections of *Partula* snails in captivity because they show great variety in their shell patterns. These geneticists have found the best conditions for keeping these snails in captivity and now several zoos have small collections in breeding groups. The pests on Moorea can be controlled using chemicals and then some *Partula* snails can be set free when the island is clear of predators and chemicals.

(a) How many species or types of snail are involved in the snail problem on Moorea? (*1*)
(b) What first upset the balance of nature on Moorea? (*1*)
(c) How did scientists try to solve this first problem? (*1*)
(d) Why did the solution go wrong? (*1*)
(e) How did the scientists then try to put things right? Did it work? (*2*)
(f) What could scientists have done before introducing Florida snails to Moorea? (*1*)
(g) Could the *Partula* snails have become extinct on Moorea? (*1*)
(h) What role can zoos and zoo scientists play in saving rare animals? (*1*)
(i) Think of one reason why the reintroduced snail population might differ from the original population. (*1*)
(j) Draw a diagram of cause and effect to show how well meaning human action had such a disastrous effect on *Partula* snails. (*6*)

*25 Amongst the stones of fast streams

Read the following passage and answer the questions.

When you lift a stone from the bed of a stream your attention is often caught by the wriggling movements of small mayfly larvae in the film of water beneath it. There are many species of mayfly and although they are not likely to be the same as those found in still water, they do share the same characteristics: three long tails and gill-fringed bodies. The fact that they remain on the stone when it is pulled out of the stream shows how well they can cling to it against the force of the current. The flattened shape of many fast-water mayfly larvae is ideal for their clinging existence.

Stonefly larvae are common in running water and resemble mayfly larvae at first glance. However they have only two tails and do not have the well-developed gills of the mayflies. Their rudimentary gills restrict them to habitats with high oxygen levels: they are not found in ponds and they are hard to keep in captivity. Certain species are found amongst the stones on the wave-washed shores of lakes, a habitat similar in many ways to fast streams.

(a) Which two characteristics do all species of mayfly share? *(2)*
(b) How are mayfly larvae adapted to fast-flowing streams? *(1)*
(c) How do stonefly larvae differ from mayfly larvae? *(2)*
(d) Why are stonefly larvae not found in ponds? *(2)*
(e) Why is a lake shore like a stream? *(1)*
(f) Suggest why many stream organisms live under stones. *(1)*
(g) List any four problems that organisms living in a small stream might face during the course of one year. *(4)*

26 Productivity

Biomass can be measured in kilograms per square metre (kg/m^2). It represents the total amount of plant and animal tissue on a square metre of ground. Primary productivity is the rate at which new plant material is made by photosynthesis. It is measured in kilograms per square metre per year (kg/m^2/y).

(a) Why is productivity measured in dry kilograms? *(1)*
(b) Look at the data below.

Area	*Primary productivity kg/m^2/y*	*Total biomass kg/m^2*
Tropical rain forest	22.00	45.00
Temperate forest (Europe)	1.25	32.00
Savanna (African grassland)	0.90	4.00
Arctic tundra	0.14	0.60
Swamp and marsh	2.00	15.00
Cultivated land	0.65	1.00
Oceans (average)	0.15	0.01

(i) Which two areas have greatest primary productivity? *(1)*
(ii) Give two explanations for the differences in productivity for the two forest areas. *(2)*
(iii) The savanna is a hot, dry place. Suggest a reason for its low productivity. *(1)*
(iv) Suggest why the productivity and biomass of cultivated land (which has fertilizers and is carefully managed) is low compared to a forest. *(1)*
(v) Swamps and marshes have 13 times the productivity of the oceans and yet they have 1500 times the biomass. Suggest two reasons for this. *(2)*
(vi) The productivity of oceans is 15 times the total biomass, whereas the productivity of tropical rain forests is half that of its total biomass. How do you account for these differences? *(2)*

27 Pest control

Read the passage carefully before you answer the questions.

Farmers and market gardeners have to control pests on their crops, but they are also under pressure from environmentalists to take care with chemical sprays. New types of non-chemical pest control are being invented and sometimes old methods can be improved. Some of the non-chemical methods used now are:

Pest scarers. These are used to frighten off crows and pigeons. They look like people but are really life-sized models dressed in old clothes. They are scarecrows and have been used for many years. The trouble with scarecrows is that the birds soon learn to ignore them because although scarecrows look like humans, they do not behave like humans. They do not move about or make a noise. So the pigeons return to eating the peas and bullfinches carry on eating soft fruit.

Another approach is to have some noisy bird scarers as well as a scarecrow. Explosive devices make a bang like a gunshot. These work quite well but they are a public nuisance and neighbours complain about the noise. They are unacceptable near houses. Some farmers use a brightly coloured triangular spinner fixed on a tall pole. As this 'top' spins, its changing appearance scares some birds. Most small birds will fly away but the pigeons will be braver. Pigeon problems can be solved by smearing wobbly jelly on the ledges where they roost. The pigeons slide about on the ledge, feel insecure and fly away. They can also be driven into nets and destroyed.

Physical protection of ripe crops. Birds can be kept away from fruit buds and fruit by fine green plastic netting. Raspberries can be protected by nets spread over metal frames. There are disadvantages with nets. Small harmless birds and hedgehogs can become caught up in them and unless they are rescued, they will die of fright or starvation.

When grain crops are gathered in, they make a fine food supply for rodents, usually rats or mice. These animals can nibble through tough grain bins made from wood or plastic, so grain is best stored in metal bins or silos.

Biological methods of pest control. These have the advantage that they can usually be directed towards the pest and nothing else. Humans use dogs to catch foxes and cats to chase birds and mice. By studying other living things, we have discovered that ladybirds will eat the greenfly on our roses. We have to study the animals we use in this way in case we make a mistake by not knowing enough. Ladybirds would be no use as greenfly controllers if they also attacked the bees.

(a) Why must farmers and market gardeners think about non-chemical ways of pest control? *(1)*

(b) If you were going to grow strawberries and raspberries how would you protect them from small birds? Would this method also protect the strawberries from slugs? *(2)*

(c) Write five lines to explain which methods of control work on pigeons and which do not. *(2)*

(d) Look at the picture of the old barn. How would the stone mushroom shaped supports discourage mice? *(1)*

(e) Design an improved scarecrow. What would be the improvements to make? How would you test your scarecrow design? *(2)*

(f) (i) Evaluate the pest control designs shown in the diagrams according to how much human labour is involved in producing and using them.
(ii) Estimate their expense and effectiveness. *(2)*

(g) Make a chart of these devices to show how expensive they are to get, how long they might last and how much of the crop each one will save. Say which ones you would use for your farm. *(5)*

(h) Does biological pest control always work? Try to find out about an attempt at biological control that did not work. *(5)*

Non-chemical pest control devices

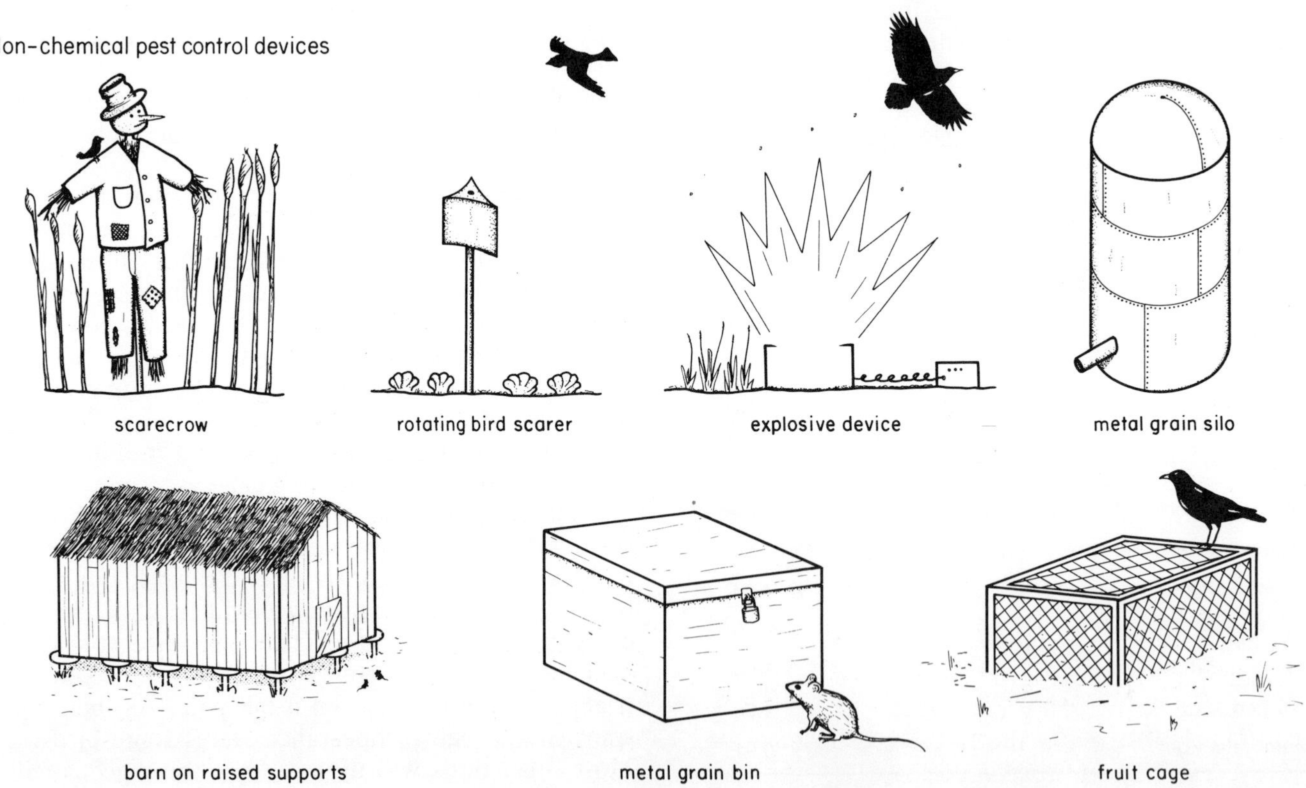

2 Problem solving

1

Firewood is an important source of energy in Kenya. Kikuyu women are often seen carrying huge bundles of firewood.

Kikuyu woman

Trains and some industries also burn wood. At the moment 75 per cent of Kenya's energy comes from fuelwood. However most of the land between 2500 m and 3000 m is already softwood forest managed by the forestry commission and used to provide fuelwood. The population is increasing at about 3 per cent per year. This means that the number of people in the country can be expected to double every twenty-five years.

There will soon be a fuelwood problem. Which of the following would NOT help to solve the problem?

A. Setting up a research station where different tree varieties and their fuelwood productivity can be studied.
B. Growing twice as many trees per hectare.
C. Giving each tree 100 g of fertilizer each year as before but in two 50 g doses every six months.
D. Cutting more gaps in the forest to prevent forest fires and improve access.

*2

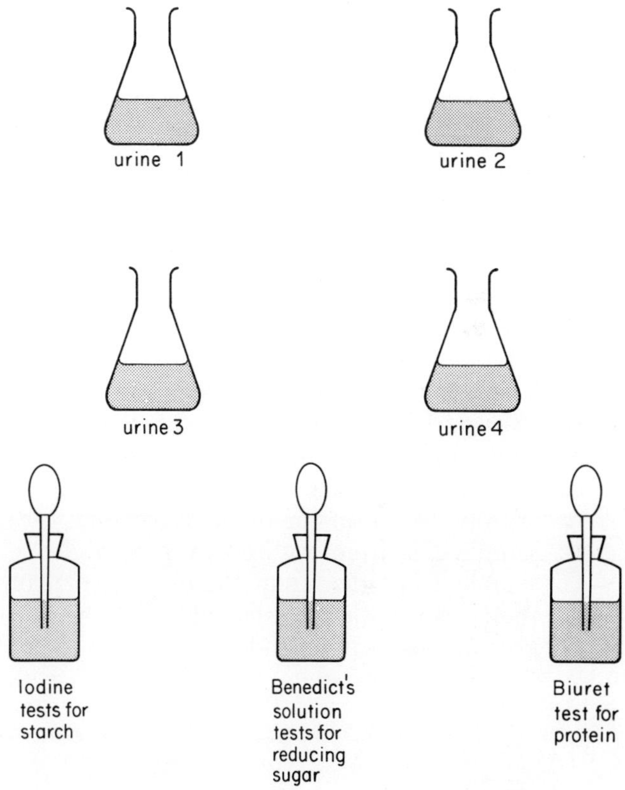

Above are shown four urine samples, 1, 2, 3 and 4. Each is from a different person:

a diabetic (diabetes mellitus)
a person with a urinary infection
a person with nephritis
a healthy person.

Diabetics who are untreated or undiagnosed have glucose in their urine. Urinary infections can cause protein, glucose and pus cells to appear in the urine. Nephritis causes protein to appear in the urine. Healthy people have urine which is glucose free, protein free and sterile.

(a) By using a microscope to look at the samples you could tell which sample came from
A. the diabetic.
B. the person with a urinary infection.
C. the person with nephritis.
D. the healthy person.

(b) Which of the following tests would be the best for finding out which urine sample came from each person?
A. Benedict's solution.
B. Benedict's solution and the Biuret test.
C. Iodine.
D. Benedict's solution, the Biuret test and microscopy.

(c) A pupil tested each of the samples in turn with Benedict's solution and was surprised when they all turned green and then orange, which is a positive result. Which of the following could be the source of error?
A. The pupil used the same pipette for sampling each of the specimens.
B. The pupil used a new boiling tube for each separate test.
C. The pupil used the Benedict's solution and forgot to add the urine.
D. The pupil did not use a timer.

(d) Another pupil tested sample 2 with Benedict's solution and it was positive. That pupil could conclude correctly that
A. sample 2 is from the diabetic person.
B. sample 2 is from the healthy person.
C. sample 2 is either from the diabetic or from the person with the infected urine.
D. no conclusion is possible yet.

A

3 Burns

People who are badly burned lose a lot of fluid and can become very ill. If they are not treated they can die. The reason for this is that burned skin oozes fluid. It cannot control fluid loss as healthy skin does. Doctors treat these patients by replacing the fluid. Plasma is given to the patient via a 'drip' into a vein in their arm. It is very important to give the correct amount of fluid. This is calculated from the following formula:

$$\text{Fluid replacement (cm}^3\text{) per period (see table)} = \frac{\text{\% of body area burned} \times \text{body mass (kg)}}{2}$$

Use the information below to work out how much fluid you would use to help a burned person.

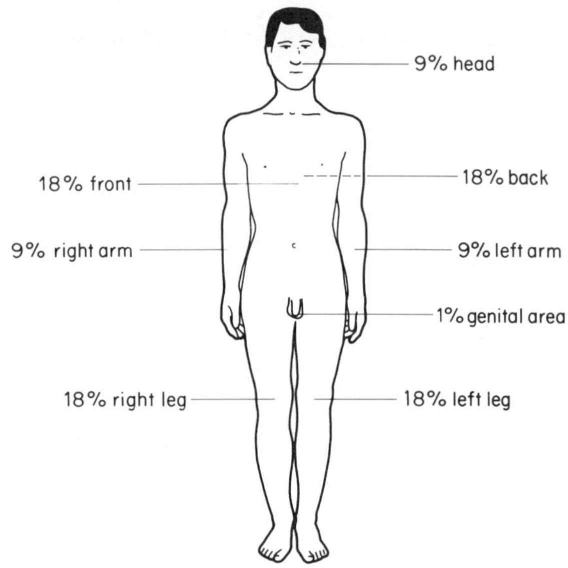

The first 36 hours are critical for fluid replacement and this is divided into set time periods:

Period 1	4 hours
Period 2	4 hours
Period 3	4 hours
Period 4	6 hours
Period 5	6 hours
Period 6	12 hours
Total = 36 hours	

(a) A patient is admitted with the left arm and the front of the body burned. The percentage burns would be
A. 1%.
B. 9%.
C. 27%.
D. 36%.

(b) A patient is admitted to hospital with both legs burned. How much fluid should you give in period 1 if the patient's mass is 70 kg?
A. 315 cm^3 per 4 hours.
B. 630 cm^3 per 4 hours.
C. 1260 cm^3 per 4 hours.
D. 2520 cm^3 per 4 hours.

(c) How much fluid will the same patient have been given in the first 24 hours?
A. 1.26 litres.
B. 1.95 litres.
C. 5.04 litres.
D. 6.30 litres.

4

The diagram shows the effects of fish farming salmon in a Scottish sea loch.

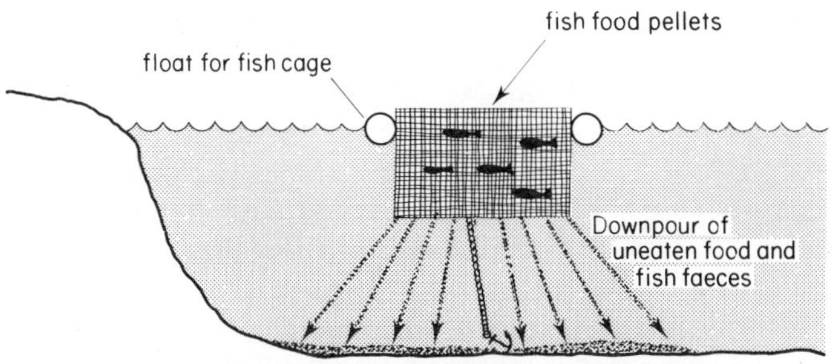

A zone of loch bed is polluted beneath the anchored fish cages. The pollution is from food, which was uneaten, and fish faeces. The scientists studying the problem called this polluting mixture of food and faeces 'Rancid Rain'. Which of these suggestions may help to solve the 'Rancid Rain' problem?

A. Dropping lime into the loch water below the cages.
B. Placing an oxygen bubbler on the loch bed below the cages.
C. Measuring the food given to the fish more accurately so the fish eat all the food.
D. Increasing the fish numbers outside the cages.

5 Cholesterol

(a) (i) Which of these activities is most likely to cause the problem of heart disease?
A. Using lipstick.
B. Smoking cigarettes.
C. Jogging.
D. Kissing.

(ii) In which of these occupations is a person most likely to suffer from heart disease?
A. Commercial traveller, sitting in a car all day.
B. Roof tiler, going up and down ladders all day.
C. Refuse collector, emptying bins all day.
D. Postman, walking about all day.

(iii) Which one of these leisure activities is most likely to reduce the chances of heart disease?
A. Drinking beer.
B. Playing golf.
C. Painting pictures.
D. Watching TV.

(b) There are many different types of fats and oils. Those classified as saturated have molecules which cannot accept any more atoms. Molecules of those classified unsaturated or poly-unsaturated have room for extra atoms. The triangular graph shows the percentage of saturated fat and poly-unsaturated oil in the fats and oils of some common foods.

The (dotted) line for margarine shows how to use this table.
Margarine contains 40% saturated fats
6% polyunsaturated oils
54% other types of oils

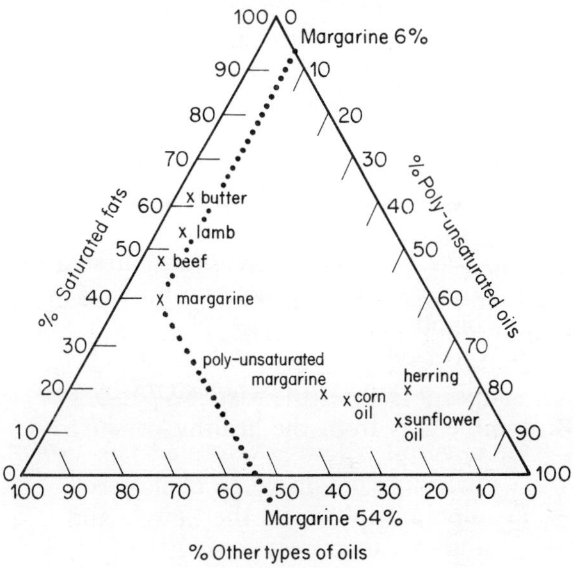

Cholesterol is a steroid which is made in the liver. It can also be present in foods which are of animal origin such as meat and milk. Foods which contain a lot of saturated fat are usually also high in cholesterol. Cholesterol production by the liver can be blocked by the use of enzyme inhibitors such as reductase. The presence of poly-unsaturated oils lowers the cholesterol content of the blood.

A ten year survey of 3806 middle aged men in the USA showed that by lowering the cholesterol level in the blood you can reduce the risk of heart disease.

(i) Which of these foods would be most likely to increase the risk of heart disease?
A. Herring.
B. Lamb.
C. Butter.
D. Margarine.

(ii) Which of these would be best to use for frying food to reduce the risk of heart disease?
  A. Sunflower oil.
  B. Butter.
  C. Poly-unsaturated margarine.
  D. Corn oil.

(iii) Which of these is a correct statement about why doctors would prescribe reductase to reduce the risk of heart disease?
  A. Reductase gets rid of cholesterol which is already in the blood.
  B. Reductase prevents a person's own body making cholesterol.
  C. Reductase prevents the cholesterol eaten in food entering the blood.
  D. Reductase strengthens the heart.

(iv) People in some Mediterranean countries use olive oil in many different ways. They fry their chips in it, put it on their lettuce and pour it into their ear to dissolve the wax. They say that as far as their heart is concerned, it helps to overcome the problems caused by their other bad habits. The best explanation is
  A. olive oil is mostly poly-unsaturated oil and attacks the cholesterol in the blood.
  B. olive oil contains enzymes which block the production of cholesterol by the liver.
  C. olive oil is low in saturated fats so it will not contain much cholesterol.
  D. olive oil lubricates the bowels and reduces the stress caused by constipation.

6

The salmon (*Salmo salar*) has an interesting life cycle. The adult salmon return each year from their feeding grounds close to Greenland, to the river of their birth. When they return to this river they swim up it to reproduce. Having entered the river via the estuary, they swim up as far as the well oxygenated headwaters. In the gravel the female digs a shallow hole (called a redd) with her tail. Here she lays the eggs which are fertilised by the male.

The eggs hatch into small fish called salmon parr. The parr remain in the river for 2 to 3 years, until they develop silvery scales. They are now called smolts and these little fish swim down the river to the sea. They head for the feeding grounds where they grow quickly into adult salmon. The cycle then repeats itself.

People have noticed a decline in the numbers of salmon returning to many salmon rivers. This could possibly be a natural phenomenon (called the cyclical returns hypothesis—which says that the number of salmon returning increase and decrease on a natural cycle). Alternatively it could be due to human interference. Which of the solutions below would be most likely to increase the numbers of returning salmon?

A. Increase the numbers of salmon parr in the river by stocking.
B. Increase the number of water bailiffs to guard the redds from poachers.

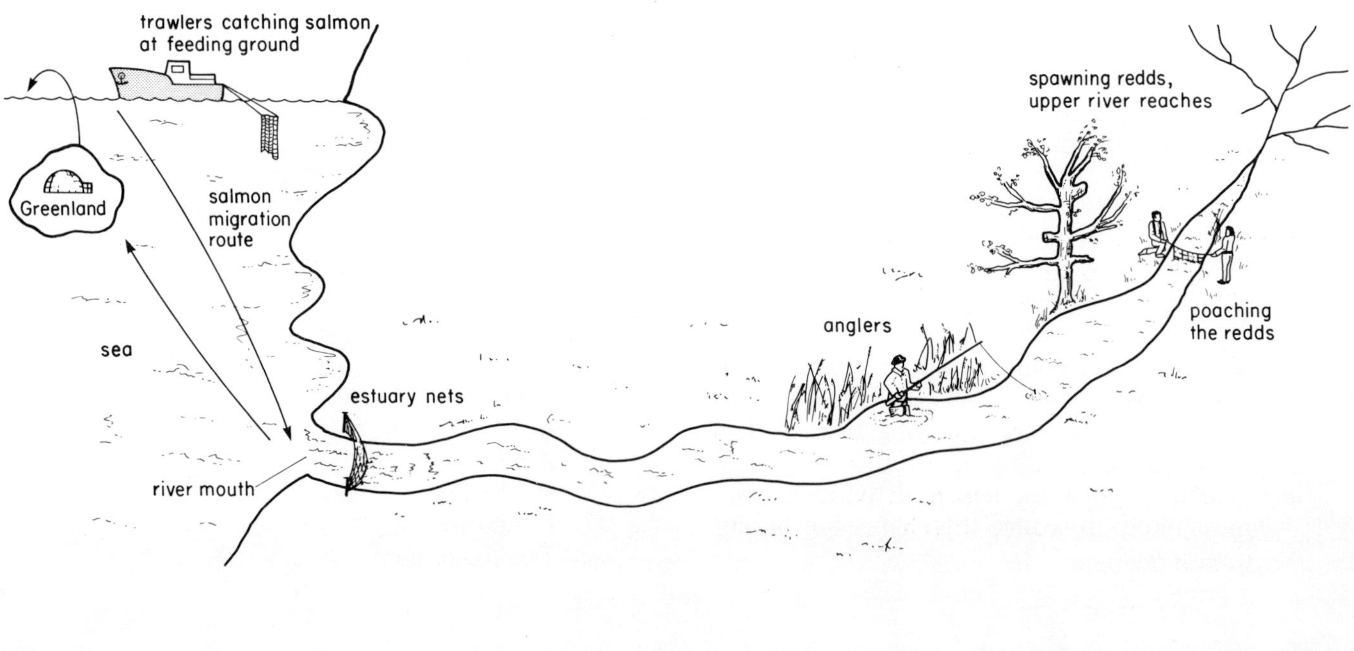

C. Allow the estuary nets to be in position only 4 days out of 7.
D. Reduce the numbers of salmon caught at the feeding grounds and ban estuary netting at weekends.

*7 Amoeba goes to Bournemouth

The Amoeba shown in the diagram is fed up with man polluting his freshwater pond and decides to go to live in the sea.

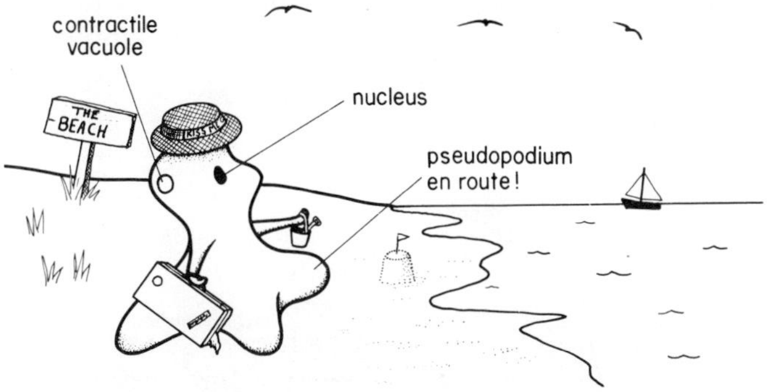

In his rush to leave his freshwater habitat for his new home, he has forgotten about altering his cell structure and function to cope with the new environment. This event is taking place in the year 2109, and by now (fortunately for Mr Amoeba) you have the technology to rebuild him! Using genetic engineering you can instruct his cell to alter its chemistry and even its structure. To help Mr Amoeba survive in his new environment you should advise him to have the following done to his cell:

A. Make a totally *impermeable* cell membrane, nothing enters or leaves the cell.
B. Remove the contractile vacuole and make the cell *isotonic* to sea water.
C. Remove the contractile vacuole and replace with an *organelle* which removes salt from the cell.
D. Give the cell a small kidney which produces urine *hypotonic* to his present cell.

Note *Impermeable*; liquid cannot pass through.
Isotonic; solutions with about the same concentrations of solutes.
Organelle; specialised part of a cell.
Hypotonic; solution which is less concentrated than another one.

*8

Some young scientists were studying alcohol production in Brazil. Sugarcane is crushed and the sugar fermented by yeast to make alcohol. One of the scientists, Paulo, says that strain B1 is the best yeast to use. There are several things that he could mean by this statement. Some of these could never be checked out scientifically.

(a) Name three things that could be checked. *(3)*
(b) Explain briefly how each of these factors could be tested. *(3)*

9

Learning skills can be a problem for some people. Three pupils investigated trial and error learning in themselves. They did a maze and counted the number of errors they made for each try. Eventually they made none. Below is a graph showing their results.

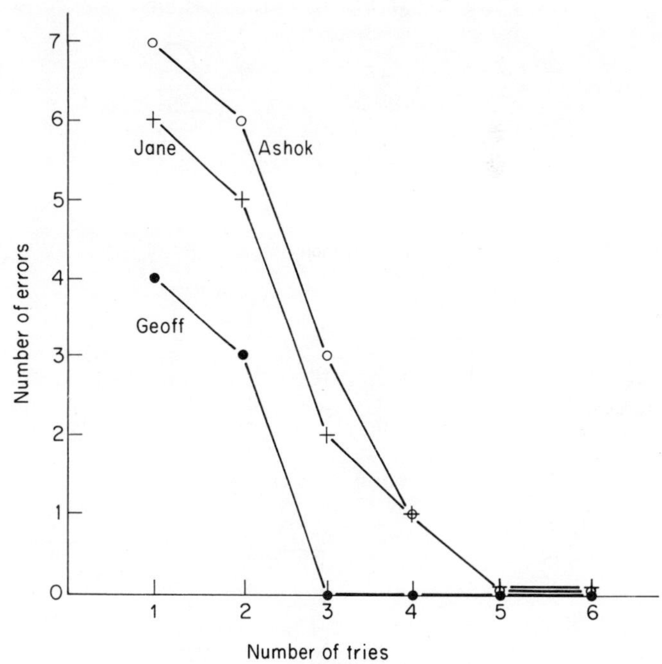

(a) Who learnt most quickly? *(1)*
(b) What does this experiment tell you about the way you should learn at school or at work? *(2)*

10

San Fun wanted to buy some flowers for her parents to celebrate the Chinese New Year. The assistant suggested she should buy a packet of 'Long Life X' powder. The assistant claimed that it would make the flowers 'live' longer if it was added to the water in a vase.
San Fun was not convinced. She decided to carry out a fair test.

Design a fair experiment that she could do to see if the assistant's claim was correct.
(Include: what to keep the same,
what to measure,
how to avoid fluke results.)

A *11

Below is a diagram of some apparatus. It is designed to measure the levels of sulphur dioxide in the air. (Sulphur dioxide is an acid gas.)

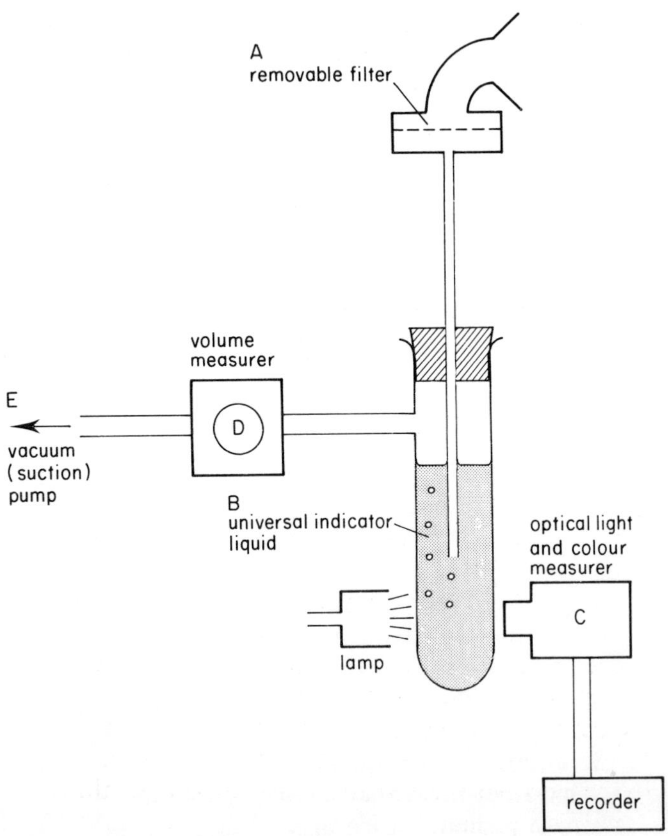

(a) Explain how the apparatus works. State a job for each part labelled A–E. *(5)*
(b) What else could be measured using this apparatus? *(1)*
(c) If you wanted to compare two areas, what two things should be kept the same if it is to be a fair test? *(2)*

12

Jon felt ill and went to see a doctor to solve his problem. She diagnosed a bacterial infection and prescribed antibiotics in the form of tablets.

The graph shows the number of bacteria and the level of antibiotic over several days in December.

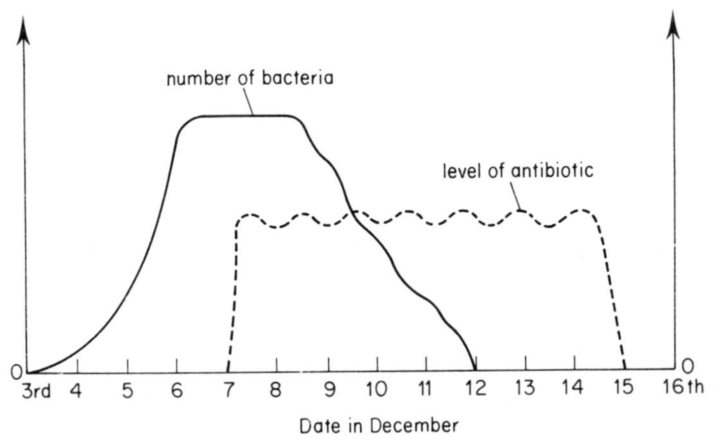

(a) On what day in December did Jon probably visit the doctor? *(1)*
(b) How long did it take before the antibiotic had any effect? *(1)*
(c) For how many days were the antibiotics prescribed? *(1)*
(d) Suggest why the level of antibiotic was not constant. *(2)*
(e) Jon was told to finish off all the tablets even though he might feel fine before the course was finished.
   (i) Why was this? *(1)*
   (ii) Draw a line on the graph to show what might have happened to the numbers of bacteria if Jon had stopped taking the antibiotics on the 11th December. *(2)*
(f) Give the name of an antibiotic. *(1)*
(g) Name a disease which cannot be cured by antibiotics. *(1)*

13

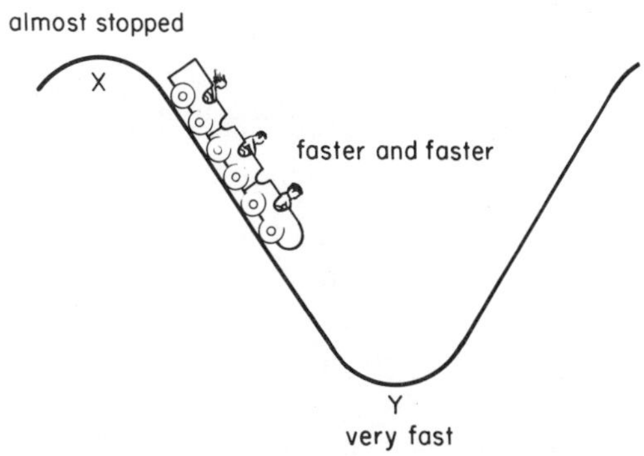

When a train on Blackpool's Big Dipper goes from X to Y it gets faster and faster. There is an acceleration of about 7 m/s^2. As your body accelerates downwards, the organs in your abdomen rise. Your stomach is squashed up against your diaphragm. This gives you 'butterflies in your tummy'. It can cause passengers to vomit on the ride.

(a) What advice might the operator give to people before buying tickets? Explain your answer. (You may wish to refer to peristalsis and the pyloric sphincter muscle.) *(3)*

As the train goes from X to Y, the blood rushes to your head. When spaceships are being launched or brought back to earth, accelerations of 80 m/s^2 are possible. This causes blood to drain from some parts of an astronaut's body and congest in others. If the brain is short of blood, the astronaut may lose vision and become unconscious.

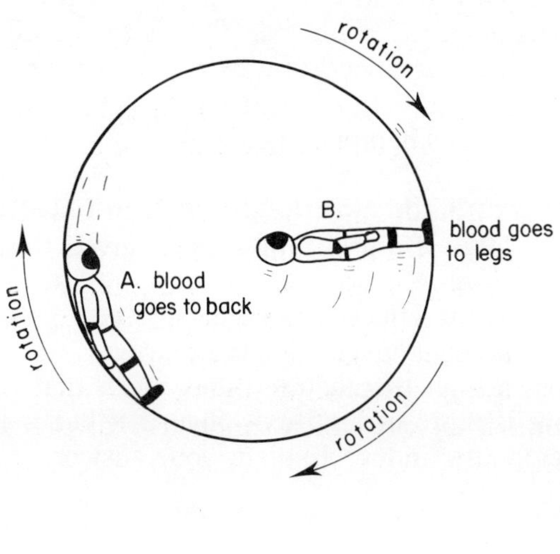

Tests have been done in centrifuges the size of people (like big spin dryers). In position A astronauts can cope with an acceleration of 150 m/s^2 for a few minutes. In position B they can cope with no more than 60 m/s^2 for the same time.

(b) An astronaut is strapped to an adjustable seat. Redraw the diagrams below to show the best position for take-off and landing. Explain why you have chosen these positions. *(6)*

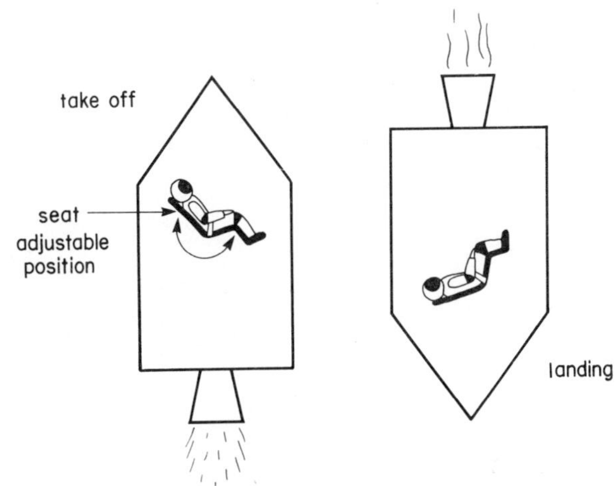

(c) Sometimes jet planes rise fast and turn corners sharply. Pilots may wear pressure suits on the lower half of their bodies. Their legs may be at twice atmospheric pressure but their head and chest are at normal pressure. Explain how this solves a problem. *(3)*

*14

Abdul is a scientist who works for a shoppers' magazine. He wanted to test how good certain toothbrushes were at cleaning teeth. He collected four brushes from a chemist.

They were: Hard nylon
Soft nylon
Hard bristle
Soft bristle.

This is what Abdul did. Read the following list carefully.

1. He prepared six agar plates in Petri dishes and labelled them 1–6.
2. He then cleaned his teeth after his lunch using the hard nylon brush with toothpaste.
3. After rinsing his mouth, he used a cotton bud and rubbed it over his teeth for 1 minute.
4. Quickly he smeared the cotton bud over the agar on plate 1.

5. He repeated this every day with the other brushes: plate 2–soft nylon
   plate 3–hard bristle
   plate 4–soft bristle.
6. For plate 5 he rubbed the cotton bud over his teeth after lunch, but he did not brush his teeth first.
7. Plate 6 was left as it was. Nothing was added.
8. All plates were sealed and left in an oven for two days. Now when bacteria grow on agar it turns cloudy. The more cloudy the agar is, the more bacteria there are.
9. All the plates were then examined after two days and the area of any cloudiness was measured in cm^2. Abdul then put his results in the form of a bar graph.

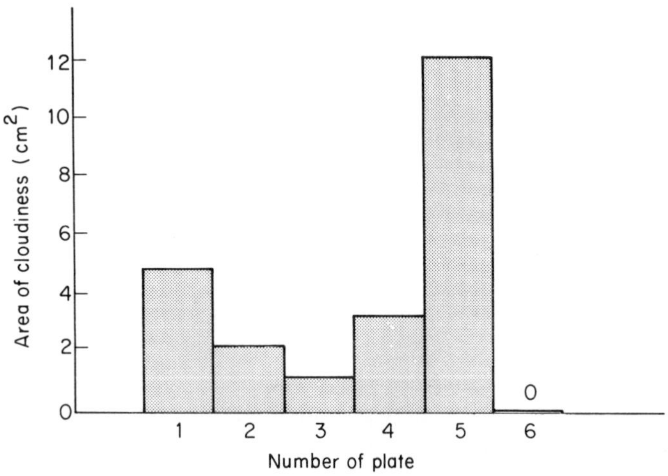

(a) Which toothbrush was the best at cleaning his teeth? (*1*)
(b) Why was plate 5 so cloudy? (*1*)
(c) Abdul should have used the same toothpaste. True or false? (*1*)
(d) When cleaning his teeth for this experiment, what must he do so that the results can be compared accurately? (*1*)
(e) Which was the worst toothbrush to use? (*1*)
(NB Human bacterial waste should not be cultured in schools.)

*15

This question concerns what happened when an epidemic disease was discovered.

Sammy was sent into hospital with an undiagnosed disease. The next day it was confirmed as typhoid. His whole family was brought into hospital for observation. Health Officers questioned them about what they had eaten and where it was eaten over the previous few days. They also checked the drinking water in their home, places of work and schools.

A tin of corned beef was suspected and sent for checking. Meanwhile the two girls next door (who went to the same school as Sammy) were taken ill. Typhoid was confirmed. 30 more school pupils entered hospital. The school was closed and the food and menus over the last few days were checked. Corned beef was on the menu three days earlier. T.V. and radio stations were told to warn people not to eat a certain brand of corned beef and it was removed from all the shelves of supermarkets and shops.

No-one else became ill for eight days. The restrictions were lifted and the school re-opened. Fortunately no-one died.

Now answer the questions.

(a) Why were Sammy's family taken into hospital even though they were apparently healthy? (*1*)
(b) Why were they asked what they had eaten? (*1*)
(c) Why was the school closed as soon as the girls were brought into hospital? (*1*)
(d) Why do you think that cooks in all schools have to keep one piece of every bit of food served every day in a fridge for 2—3 days? (*1*)
(e) Suggest a possible way in which the meat could have become infected. (2)

16 Overdose (self-poisoning)

(a) Many drugs prescribed by doctors are safe as long as they are taken in the correct dose. Drugs are tested on animals before they can be taken by humans. From tests on animals, chemists can find out how poisonous a particular drug could be. This is measured using the therapeutic index.

$$\text{Therapeutic index} = \frac{LD_{50}}{ED_{50}}$$

LD_{50} is the dose of the drug which is lethal to 50% of a group of animals given that drug.

ED_{50} is the effective dose in 50% of a group of animals given that same drug.

Does a high therapeutic index mean that the drug is more or less toxic than if it had a low therapeutic index? Explain your answer. (*3*)

(b) The liver is a very important organ. But the liver cells can be damaged by certain drugs, such as paracetamol. Paracetamol is a commonly used drug and it is safe in the correct dose. But if an overdose is taken it can fatally damage the liver cells. The liver is not damaged immediately but treatment is needed urgently if paracetamol levels are high.

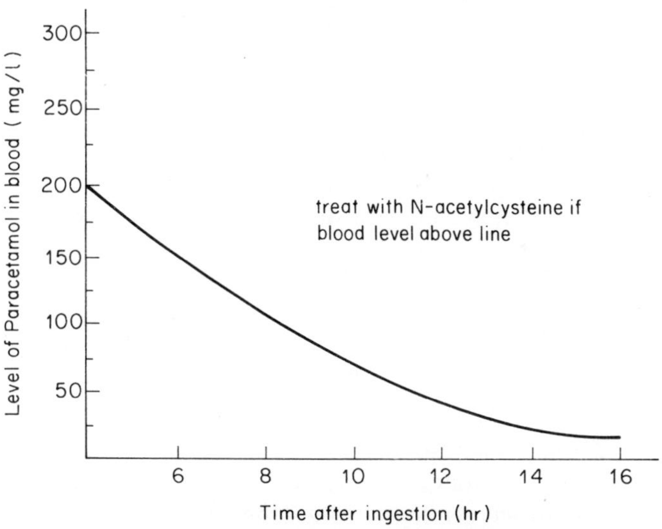

There is an 'antidote' to paracetamol. This is another drug which can stop paracetamol damaging the liver. The antidote must be given if levels of paracetamol are above the line on the graph. The antidote to paracetamol is called N-acetylcysteine. Treatment started more than 15 hours after the taking of paracetamol is ineffective.

(i) Suppose a patient is taken to hospital 8 hours after taking an overdose. The paracetamol level in that person's blood is 150 mg/l. Is treatment with the antidote necessary? Explain your answer. *(3)*

(ii) Which other drug taken regularly in our society can damage the liver if taken often and in excess? *(1)*

(iii) Find out about the functions of your liver. It is a fantastic organ. If it was damaged, in what ways would you become ill? *(5)*

17

This question is about calculating the number of bacteria in a sample of river water.

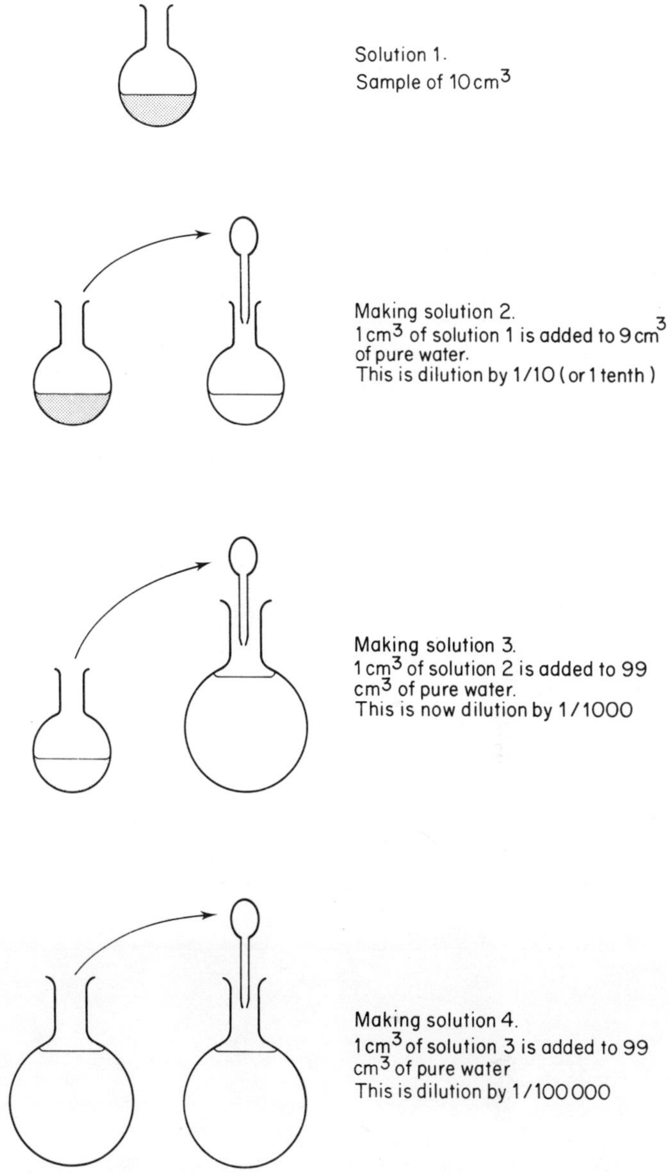

1/10 of solution 4 is grown on agar and six colonies grow.

Think about the dilutions and work out the original number of bacteria in the sample of river water.

*18 Dead log

Miriam was investigating life in a dead log. It was a cold October day when she removed a large piece of bark from the log.

This is what she saw:

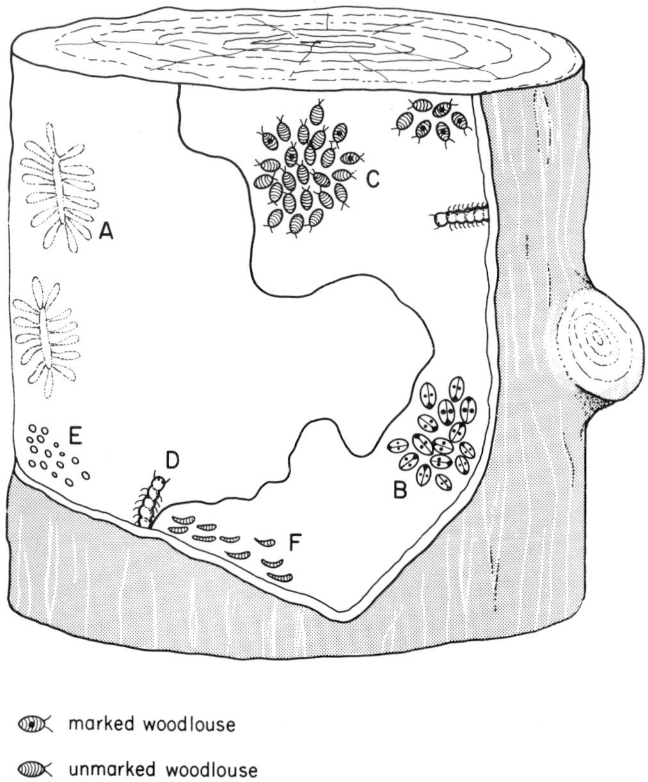

(a) **B** is a cluster of two spot ladybirds. Suggest why they were there, because they do not usually live under bark. *(2)*

(b) **C** is a crowd of woodlice. They usually live under dead tree bark. Miriam caught some and marked them with paint. Then she returned them to the log. They were found like this two days later. Had Miriam marked all the woodlice on the log? Explain your answer. *(1)*

(c) **D** is a carnivorous centipede. It lives under treebark. Close to it, at **F**, are some shells of eaten woodlice. Some of them are paint marked. Make two guesses about the behaviour of woodlice and centipedes that could be tested. (2)

(d) There are no ladybird husks close to the centipedes' lair. Suggest tests to find out whether ladybirds keep away from centipedes or centipedes keep away from ladybirds. What would you expect to find out. Explain your answer. *(4)*

(e) **A** is a pattern with a deep centre channel and smaller side channels. It was made by a large woodboring beetle. A large female beetle went under the bark to lay her eggs. How would this pattern be formed? *(2)*

(f) **E** marks a group of boreholes made by very small woodboring beetles as they left the galleries where they had lived for three years.
   (i) Pick out two differences between the patterns for the galleries of the large woodboring beetles and the small ones. What could you guess about their ways of life? *(2)*
   (ii) Explain why their ways of life differ. (2)

(g) Miriam was estimating the woodlouse population of the log. She caught some woodlice, marked them and released them again. These were her results: ()

Woodlice captured and marked with paint on Monday	50
(These were released onto the log.)	
Woodlice recaptured on Wednesday;	
marked with paint	20
unmarked	80
total recaptured	100

Use this formula to calculate the woodlouse population on the log.

$$\frac{[\text{Total marked with paint} \times \text{Total recaptured}]}{[\text{Total of marked woodlice recaptured}]}$$

(4)

19 New garden

Julia has moved house and has brought a lot of her garden plants with her. The new house is in a limestone area. At one side of the front garden the soil is only 15 cm deep above the limestone rock and soil tests give pH 8. At the other side, under trees, falling leaves and perhaps composting over many years have produced a soil of pH 5. Some parts of the garden are shaded all day. Other parts are not shaded. The following diagrams are plans showing pH levels and light/shade.

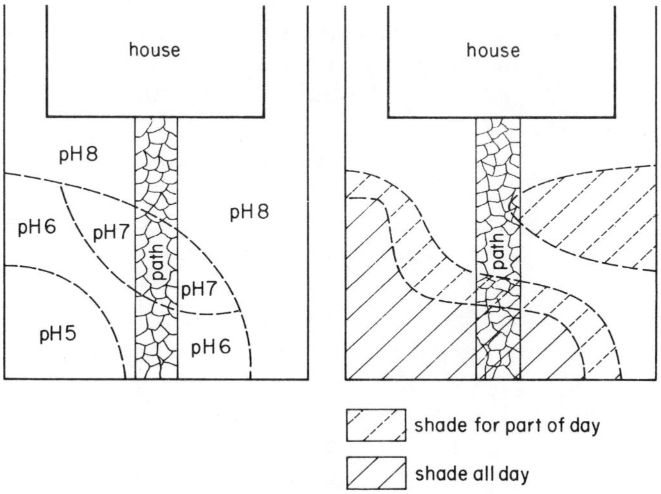

The growing conditions preferred by ten plants are listed below.

Plants needing acid soil (pH 5)
Rhododendron (R)
Magnolia (M)
Heather (H)

Plants which do well on alkaline soil (pH 8)
Catmint (CM)
Rose of Sharon (RoS)
Stonecrop (Sc)
Spanish Gorse (SG)
Periwinkle (P)
Lupin (L)
Red Hot Poker (RHP)

Plants for shade
Rhododendron
Periwinkle
Rose of Sharon

Plants for dry sunny sites
Spanish Gorse
Catmint
Rose of Sharon
Stonecrop

Plants for sites which are not too dry and not too shaded
Magnolia
Red Hot Poker
Lupin
Heather

(a) Draw a plan of the garden showing by initials (e.g. RHP for Red Hot Poker) where the plants listed could be grown. (*8*)

(b) (i) Which two plants could not easily be grown in this garden? (2)

(ii) What could be done to make it possible to grow them? (2)

3 Patterns

1

Six leaves from different plants were collected from a garden and the following key was made. Use the key to identify each of the six leaves.

KEY

1. Leaf made up of separate leaflets .... Laburnum.
   Leaf not made up of separate leaflets .... go to 2.
2. Leaf at least twice as long as broad......................... Cotoneaster.
   Leaf not twice as long as broad .... go to 3.
3. Leaf with lobes .... Hawthorn.
   Leaf not lobed .... go to 4.
4. Leaf with smooth edge. .... Lilac.
   Leaf with toothed or serrated edge .... go to 5.
5. Leaf with sharp spines on its edge.......................... Holly.
   Leaf with a serrated edge but not sharp spines......... Mock Orange.

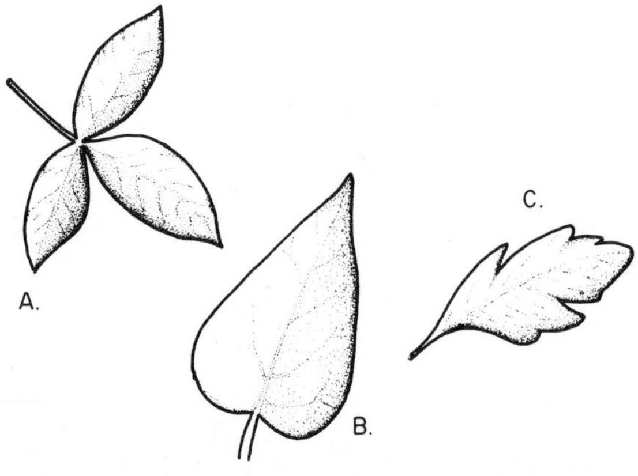

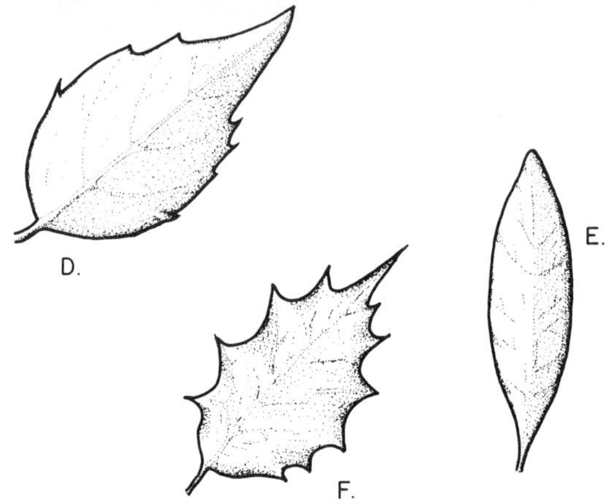

Which leaf is

(a) Cotoneaster,
(b) Hawthorn,
(c) Holly,
(d) Laburnum,
(e) Lilac,
(f) Mock Orange?

2

Diagram 1 shows a flower and an expanded floret from the pea family: it is called the sweet pea.

Diagram 1

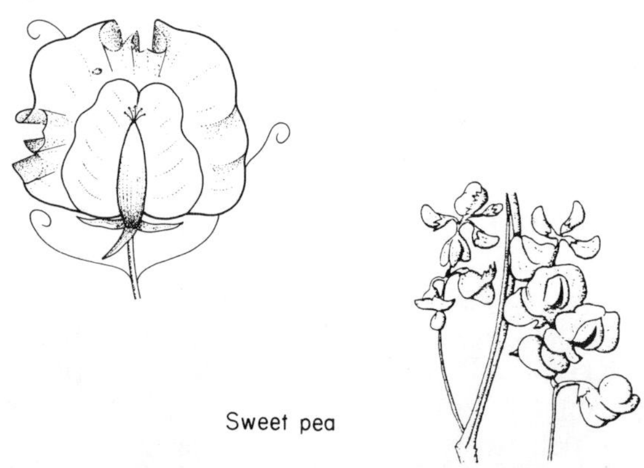

Diagram 2 shows the flowers and florets of four plants. Which is not a member of the pea family?

Diagram 2

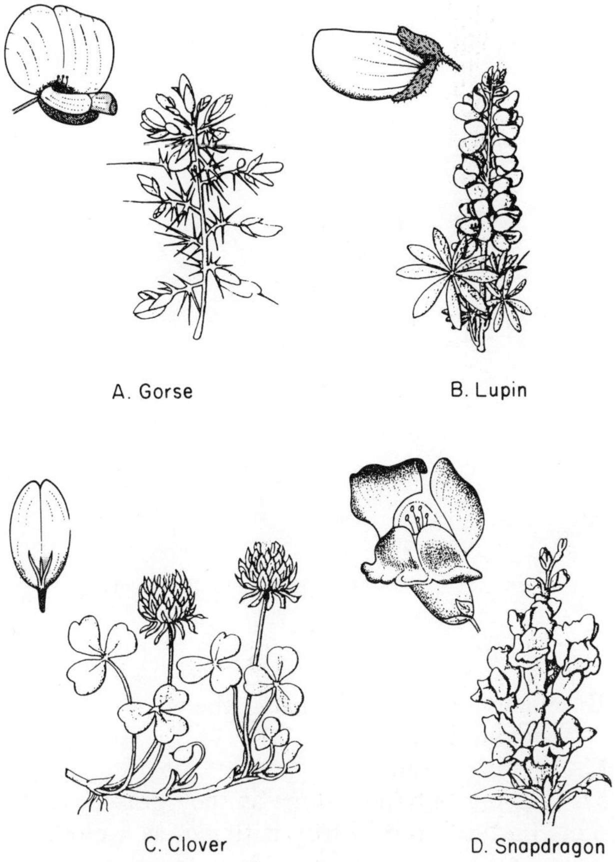

3

Diagram 1 shows a harebell. It is a member of the bellflower family. Bellflowers are usually blue.

Diagram 1

Which of the flowers in diagram 2 could be another member of the bellflower family?

Diagram 2

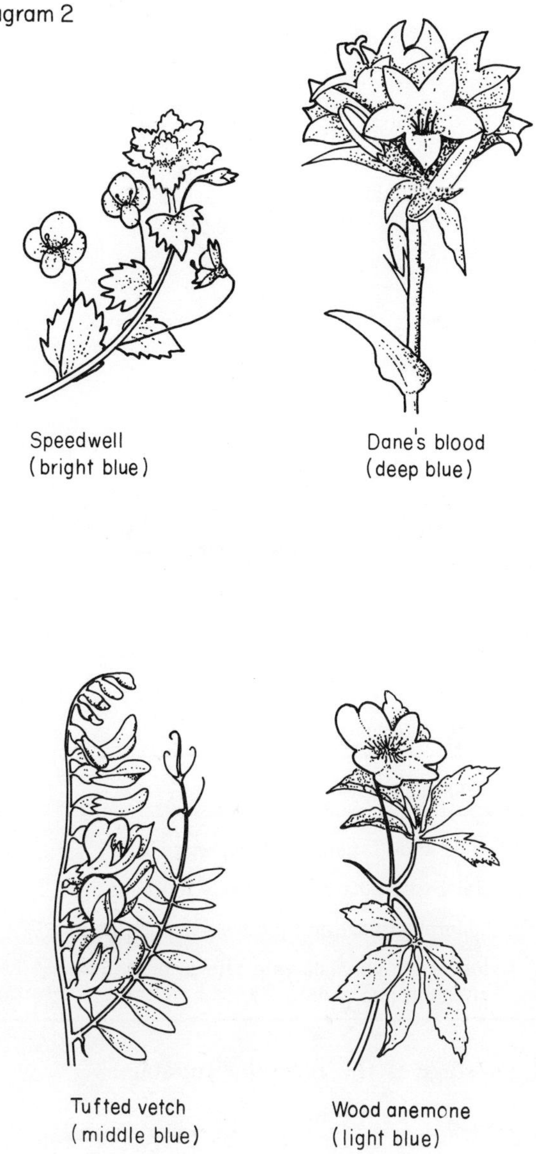

4 Butterflies

These questions are about the life history of butterflies. Butterflies develop in four stages. Each stage is different. Use the diagram overleaf to help you to answer the questions.

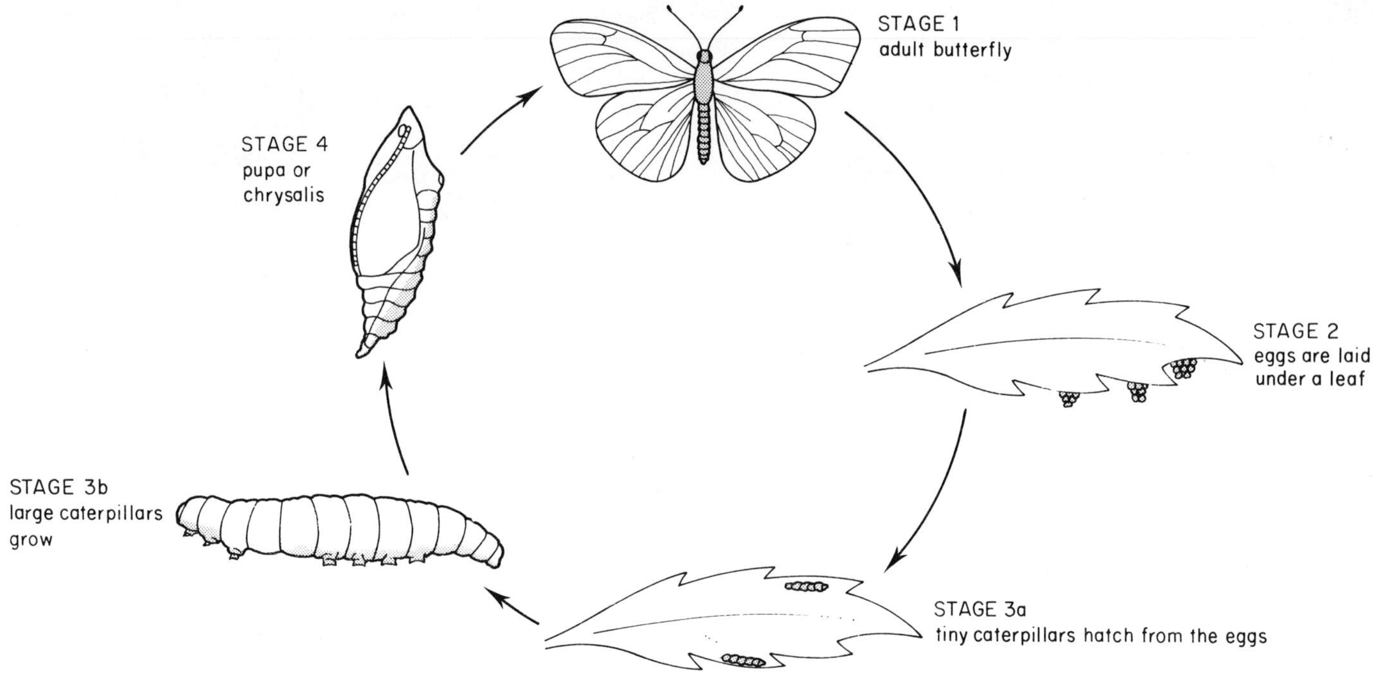

More about the stages

Stage	*What it does*
1	Flies about, mates and lays eggs.
2	Baby caterpillars grow inside the eggs.
3	Caterpillars eat and grow very much larger.
4	Inside the chrysalis, the caterpillar's body is changed into a butterfly.

(a) Which stage is the reproducing stage?
- **A.** 1.
- **B.** 2.
- **C.** 3.
- **D.** 4.

(b) When does most of the eating and growing occur?
- **A.** Stage 1.
- **B.** Stage 2.
- **C.** Stage 3.
- **D.** Stage 4.

(c) Which part of the life history shows the greatest change in appearance?
- **A.** From 3a to 3b.
- **B.** From 3b to 4.
- **C.** From 2 to 3a.
- **D.** From 3b to 1.

(d) Caterpillars shed their skins four times in stage 3. This is because they are
- **A.** growing fast and their skins become too tight.
- **B.** changing colour to match their surroundings.
- **C.** keeping clean.
- **D.** wearing out their skins as they graze.

(e) This butterfly life history is drawn as a circle because
- **A.** butterflies move in circles to spread and reproduce.
- **B.** one stage leads to the next in steps.
- **C.** there is always the same number of stages.
- **D.** the same order of stages always leads to new butterflies.

(f) The pupa stage is often the one that lives through the winter. This is due to
- **A.** a hard coldproof skin.
- **B.** the pupa not needing to feed and grow.
- **C.** the fact that there is no food.
- **D.** the death of all butterflies in winter.

(g) If the life history of a butterfly takes four weeks, how many times will it occur in the three months of summer?
- **A.** Once.
- **B.** Twice.
- **C.** Three times.
- **D.** Four times.

(h) Which two stages must be able to recognize a food plant?
- **A.** 1 and 2.
- **B.** 1 and 3.
- **C.** 2 and 3.
- **D.** 3 and 4.

5 Curing leukaemia

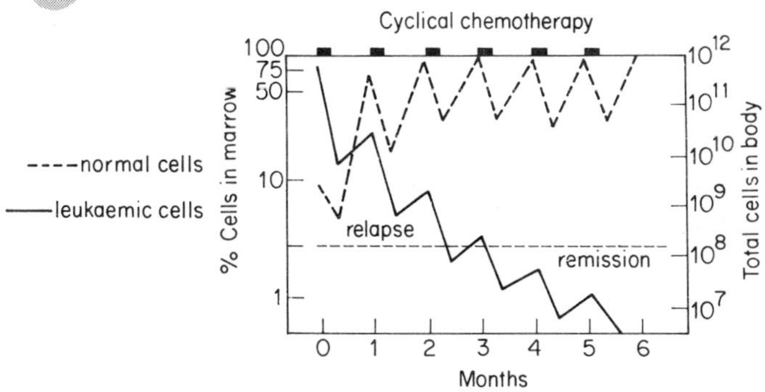

Source: *Essential Haematology*. Hoffbrand and Pettit 1984 (Blackwell Scientific Publications Ltd.)

Notes on leukaemia:
Leukaemia cells divide like ordinary cells.
Leukaemia cells can be recognised only when at least 4% of cells in bone marrow are of this type.
When leukaemia cells are no longer seen in bone marrow, the patient is said to be in remission (i.e. disease free).

(a) In which month did the patient finally go into remission?
A. Month 1.
B. Month 2.
C. Month 3.
D. Month 4.

(b) From the graph above it can be seen that during a block of chemotherapy
A. only leukaemia cells decrease in number.
B. only normal cells decrease in number.
C. both normal and leukaemia cells decrease in number.
D. both normal and leukaemia cells increase in number.

(c) Chemotherapy is given for one week followed by no treatment for three weeks to allow
A. the normal cells to recover at a greater rate than leukaemia cells.
B. the leukaemia cells to recover at a greater rate than normal cells.
C. less expensive treatment.
D. the percentage of leukaemia cells to increase.

6 Electrocardiogram

The diagrams are three different traces of differing heart rates and rhythm.

Diagram 1

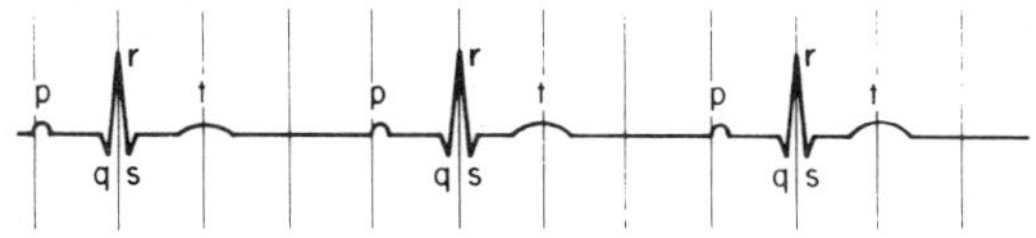

Above is a trace of a normal electrocardiogram (ECG) pattern. One complete heart cycle causes the tracings of p, q, r, s and t patterns on the graph paper. Each time the heart beats one complete cycle the pattern repeats. Diagram 1 shows three complete cycles.

Contraction of the atria causes the ECG to trace the p pattern. The q, r, s pattern on the tracing is due to the ventricular contraction. The t pattern on the tracing is due to the ventricles returning to resting state (ready for the next cycle).
All ECG traces are recorded on graph paper.
One large square = 0.2 seconds.
Therefore
5 large squares = 1 second
300 large squares = 60 seconds = 1 minute.
The patient's heart rate therefore is the r to r interval.
Diagram 2 summarizes the timings.

Diagram 2

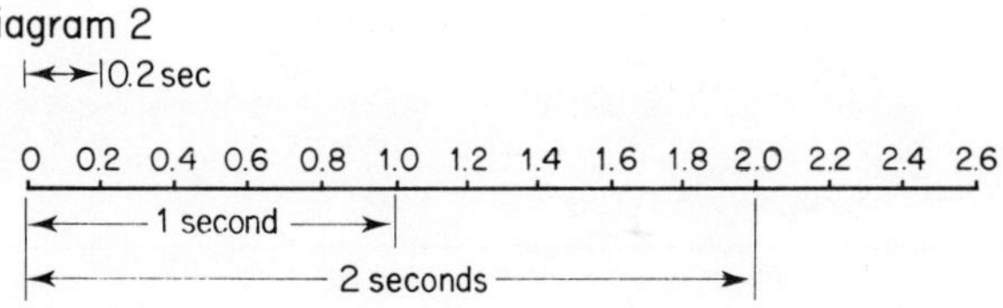

Diagram 3

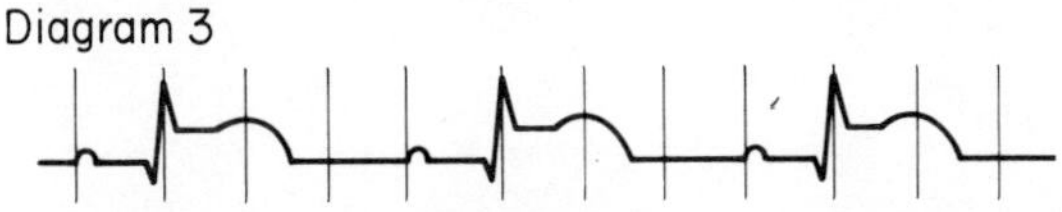

(a) (i) This patient's ECG is not normal because
A. the p pattern is abnormal.
B. the s, t pattern is abnormal.
C. the q pattern is abnormal.
D. the q, r, s pattern is all abnormal.

(ii) The patient has an abnormality in
A. the atria.
B. both atria and both ventricles.
C. the ventricles.
D. the left ventricle.

Diagram 4

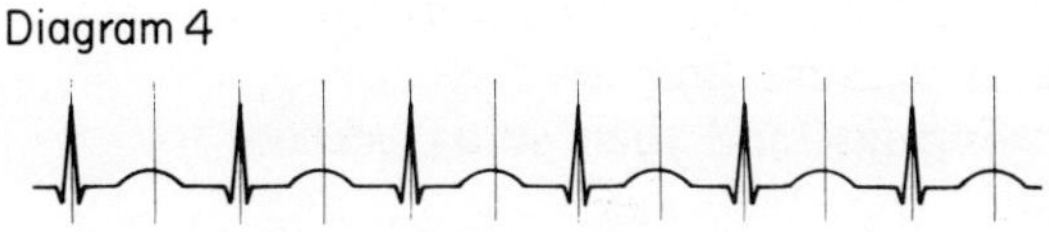

(b) (i) This patient has a heart rate (a pulse) of
A. 300 per minute.
B. 150 per minute.
C. 75 per minute.
D. 50 per minute.
(ii) Missing from diagram 4 are
A. the p patterns.
B. the q patterns.
C. the r patterns.
D. the s patterns.

Diagram 5

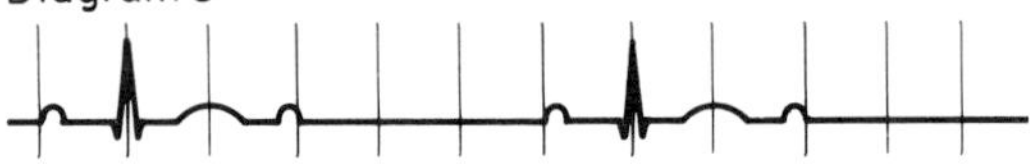

(c) This patient has a heart rate (a pulse) of
A. 300 per minute.
B. 150 per minute.
C. 75 per minute.
D. 50 per minute.
(d) Which patient, or patients, would you expect to have a fast beating heart?
A. The patient in a.
B. The patient in b.
C. The patient in c.
D. The patients in b and c.

7

The diagrams below show a drone and a queen bee.

(a) Write down three similarities between the two. *(3)*
(b) Write down three differences between the two. *(3)*

8

Below is a simplified foodweb of a pond.

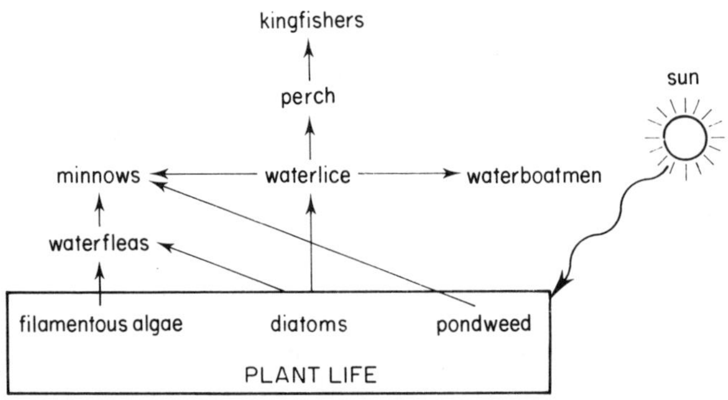

(a) Name the producers. *(3)*
(b) Which one is an omnivore? *(1)*
(c) Which ones are carnivores? *(3)*
(d) Write down any two food chains in the web above. *(4)*
(e) What would happen initially to the fish and producers if the kingfishers were removed? (2)

9 Childhood illness

Every boy and girl in the fourth year of a school were asked what diseases they had had. The results are below.

Disease	*Boys*	*Girls*	*Total*
Cold	100	100	
'Flu'	45	55	
Measles	16	14	
German Measles	17	3	
Typhoid	0	0	
Chicken Pox	27	33	
Whooping Cough	1	1	
Athlete's Foot	30	10	

The number of boys is equal to the number of girls. The total is 200.

(a) Which disease is the most common? *(1)*
(b) Why is this the commonest? *(1)*
(c) Which two diseases are the rarest? *(2)*
(d) Suggest a reason for this. *(1)*
(e) Do you think that this pattern of diseases would be similar throughout the rest of the school. Give a reason. *(2)*
(f) Due to a poor option system in this particular school, the boys in the fourth year do much more games and PE than the girls.
Looking carefully at the figures, can you see what effect this had on the number for a particular disease. Explain your answer. *(3)*

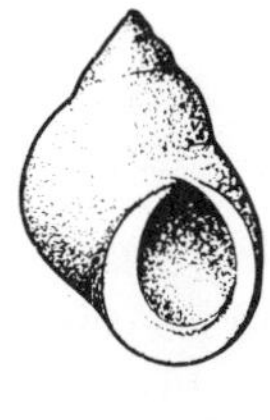
periwinkle

10 Shore feeders

Look at the organisms found on a rocky shore and their diet.

Dogwhelks feed on mussels and barnacles.
Crabs feed on dogwhelks, periwinkles and mussels.
Barnacles feed on animal and plant plankton.
Mussels feed on animal and plant plankton.
Periwinkles feed on seaweed.
Animal plankton feed on plant plankton.
Seaweeds and plant plankton are plants.

Draw a foodweb using the above information.

crab

barnacle

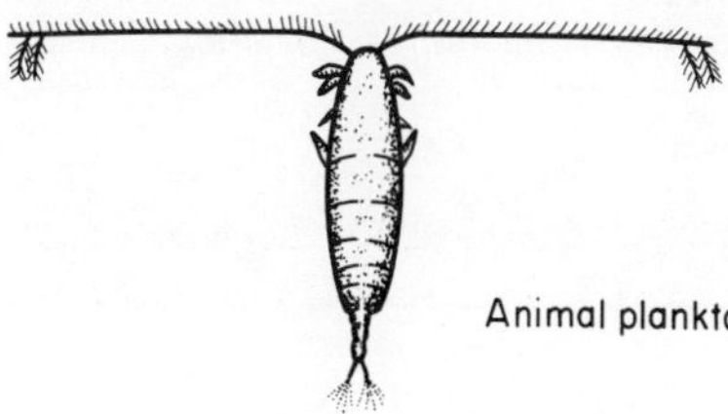
Animal plankton

mussel

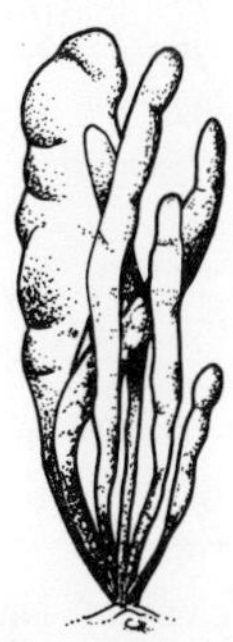
Sea weed

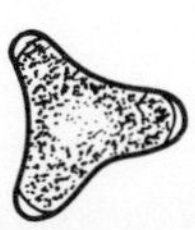
Plant plankton

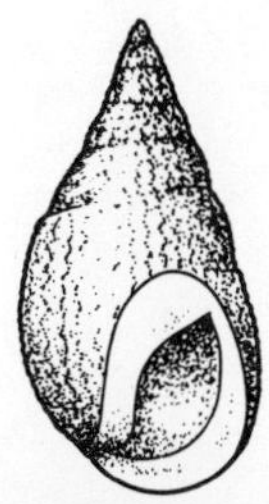
dogwhelk

11 Snowtracks

Some children were walking on a fine, sunny, snowy day by the side of a wood. They found some tracks and they tried to work out what the animals might have been doing. Use this key to help you.

(a) Which four animals had passed over the snow? *(4)*

(b) Two of the animals were going after the acorns. Which ones were they? *(2)*

(c) Which animal left hoof marks? *(1)*

(d) Two of the animals moved in a series of leaps. Which two were they? *(2)*

(e) Which animal sat down in the snow? *(1)*

(f) Work out the order in which the tracks were left. How did you work out your answer? *(3)*

12 Questionnaire results

An opinion poll interviewed several people in July 1986 to find out what they knew about biology. People were classified by sex or by their educational background ie none, CSE level, 'O' level, 'A' level or D/D (degree/diploma).

Educational background was also classified into science and non-science.
The numbers represent the total percentage of people giving that particular response.
The question asked was: "Do you think that oxygen is transported to the cells of the human body by white blood cells; hormones; enzymes; red blood cells?

	Response					
	White blood cells	*Hormones*	*Enzymes*	*Red blood cells*	*None of these*	*Don't know*
All	16	2	4	62	0	17
Male	19	2	3	61	1	15
Female	13	1	5	62	0	18
Science level						
None	18	2	4	53	0	22
CSE	25	1	3	56	1	14
O level	12	1	4	77	1	5
A level	13	0	2	79	0	6
D/D	4	0	3	87	0	6
Non-science level						
None	16	2	3	54	0	24
CSE	23	2	5	59	0	10
O level	17	1	4	69	1	8
A level	9	2	5	75	0	9
D/D	14	0	2	79	0	5

Source: *Journal of Biological Education* (1987, Vol 21, No. (1))

The correct answer is of course red blood cells!

(a) What pattern do you notice about the percentage of people giving the correct answer and their educational background? (2)
(b) What effect does the science background (level) have? (1)
(c) Note down any two other patterns you notice in the figures. (2)

13

Construct a key for the identification of the following six animals:

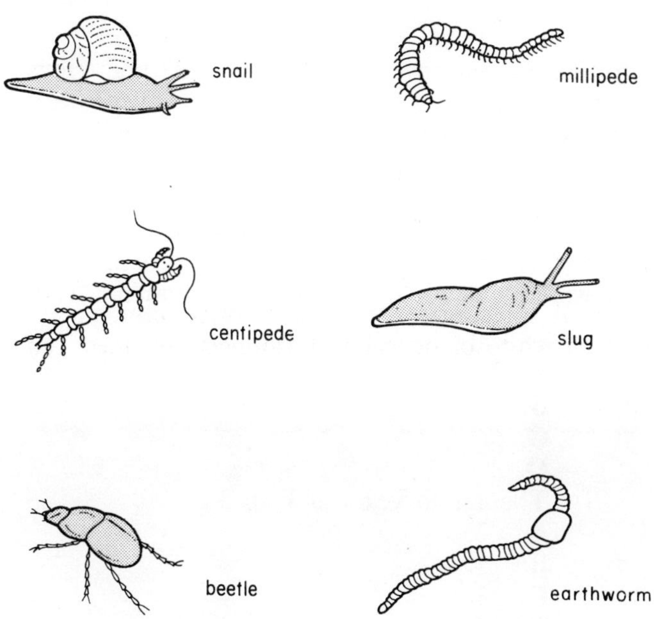

14 Eggs

This question is about an experiment with frogs' eggs. The experiment was to see if there is a connection between the age of a donor nucleus from one frog's egg and the percentage survival rates of the resulting frog embryos.
This is what was done:

The fertilised eggs from two different parent frogs were used. At a range of times after fertilisation the nucleus was removed from one egg in group A. This nucleus was replaced with a nucleus from a frog's egg in group B. The new egg was allowed to develop as normal.

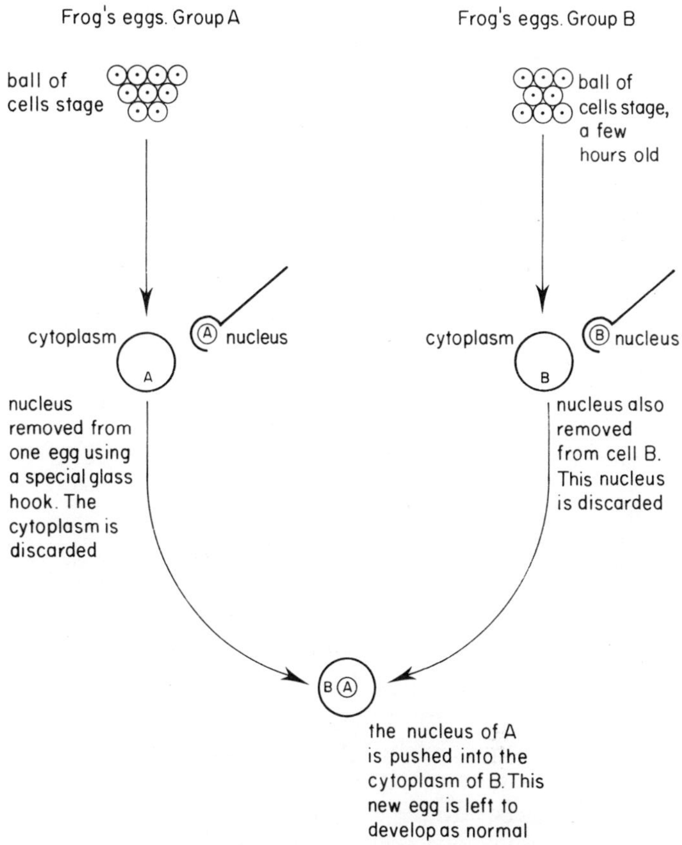

Age of donor nucleus (hours)	*% surviving recipient embryos*		
	24 h after transplant	*72 h after transplant*	*144 h after transplant*
24	100	95	85
40	95	75	25
60	35	15	5
80	15	5	0

(100% means that all recipient embryos survived)

(a) State clearly two patterns or trends shown by these results. (2)
(b) Suggest an explanation for one of the patterns you state. (1)

15 Muscle power

Some pupils were investigating how strong their muscles were by pushing as hard as they could on a newton (N) meter scale. (Like a bathroom scale but with newtons instead of kilograms.)
Use their results to answer the questions.

Pupil's name	Muscle force (N) Triceps	Biceps	Fingers	Pectorals (chest)	Thigh
Adrian	400	500	300	150	1200
Andrew	300	350	450	400	850
Darren	200	650	400	350	1000
John	500	800	435	550	1350
Khizer	400	650	400	400	1110
Mark	200	300	150	250	700
Paul	330	350	250	500	900
Philip	500	800	600	500	1550
Zahid	500	700	550	435	1050
Angela	350	150	300	200	800
Helen	250	290	350	200	550
Joanne	200	200	150	200	950
Julie	300	500	500	250	1350
Michaela	150	100	200	300	1000
Sharon	200	200	150	250	800

(a) (i) Draw a scattergram (points plotted on a graph without drawing a line) to see if there is any relationship between the strength of biceps and thigh muscles. (*5*)

(ii) Is there a relationship? What does it seem to be? Refer to two pieces of information from your scattergram which support your conclusion. (*4*)

(b) Distinguish between the boys and the girls on the scattergram. What differences are shown in their plots? (*3*)

(c) Suggest one reason why there is such a range in the force from these pupils' fingers. (*1*)

(d) Draw a histogram showing the finger force of these fifteen pupils. (*5*)

(e) Find one pattern linking the biceps and triceps results that is true for all but two pupils. (*2*)

(f) Are all boys stronger than all girls? Give a reason for your answer. (*2*)

(g) How many pupils could lift a 1000N weight attached to their ankle? List these pupils in order of leg strength. (*3*)

(h) Which boy may be a weightlifter? (*1*)

(i) Which girl may be a strong swimmer? (*1*)

(j) Which girl is probably slightly built? (*1*)

(k) Which boy do you think has the best muscles? How did you decide? (*2*)

16 Fish scales

The two accompanying diagrams show scales from two different fish. Both are at the same magnification.

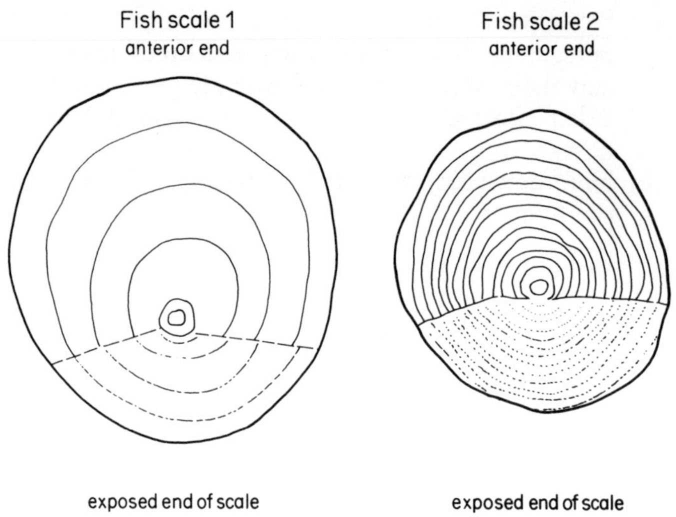

As fish scales grow during a fish's lifetime, rings of a bony substance form around a central point. The new ring is laid down around each existing ring of the scale. Each winter the amount of growth slows and this shows up on the scale as a dark, narrow band. (These are seen most clearly on the anterior end of the scale.) In between these dark winter rings is the amount of summer growth.

(a) (i) The age in years of Fish 1 is
A. 2.
B. 4.
C. 6.
D. 8.

(ii) The age in years of Fish 2 is
A. 7.
B. 11.
C. 15.
D. 17.

(iii) During which years did Fish 1 and Fish 2 have similar growth rates?
A. 1 and 2.
B. 3 and 4.
C. 5 and 6.
D. Not enough data available.

(iv) Which of these statements is correct?
A. Fish 1 grows faster on average than Fish 2.
B. Fish 2 grows faster on average than Fish 1.
C. Both fish have the same average growth rate.
D. There is not enough information available to decide which fish has the faster average growth rate. (*4*)

(b) Why should summer growth be greater than winter growth? (*2*)

(c) Both Fish 1 and Fish 2 were caught in the same river. This river, throughout its whole length to the sea, is a constant environment. Fish 1 is a different species from Fish 2. What can you deduce about the life histories of the two fish which explains the difference in ring scale patterns? *(4)*

17

Look at the following diagrams of imaginary animals.

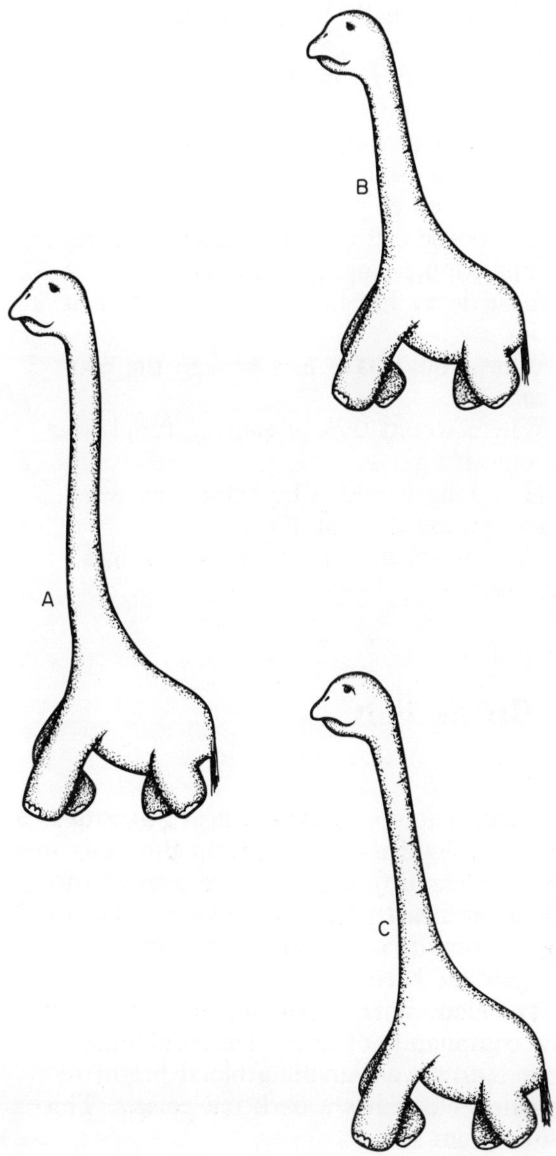

The figure below shows the blood pressure in the neck of animals A and B. The artery measured leads into the brain. The pressure was measured with the animals standing up and lying down.

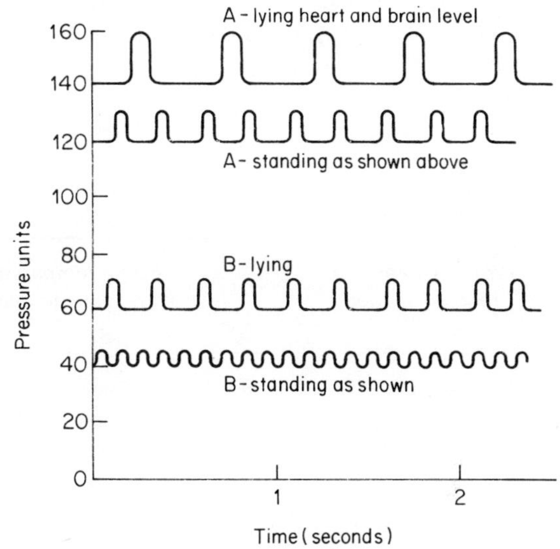

(a) Calculate the number of heart beats in beats per minute of
  (i) A lying.
  (ii) B lying. *(2)*

(b) What explanation do you have for the difference in lying and standing in the same animal? *(2)*

(c) What pattern do you see between heart beat, blood pressure and neck length? *(2)*

(d) Draw as carefully as you can the pattern you might expect for animal C both lying and standing. *(4)*

*18 Monitoring compost

A gardener wanted to find out about the temperature changes in rotting grass. It was important to him because rotting grass generates heat. This can speed up the rotting process and kill off weed seeds. He was interested to see whether the weed seeds would be killed off and when the compost might be ready. He collected a black refuse bag of grass cuttings and placed it in a sunny corner, outside. Then he fixed up a temperature probe so that it was in the centre of the grass in the bag. The probe was connected to a data logger. Temperature readings were taken at regular intervals. After two weeks, the data logger was connected to a computer. The graphics programme changed the data into a graph.

The table for the graph is given below. Draw the graph and use it to help you to answer the questions.

Days	*Probe temperature reading logged °C*	*Days*	*Probe temperature reading logged °C*
1	10 15 20 15	9	34 41 48 45
2	12 18 24 20	10	42 51 60 59
3	16 22 28 24	11	58 65 74 73
4	20 26 32 28	12	72 81 90 91
5	24 30 36 32	13	88 97 106 101
6	28 34 40 35	14	96 101 104 97
7	30 35 39 33	15	90 93 96 89
8	28 34 40 37		

(a) (i) What was the time interval between temperature readings? *(1)*
(ii) What was the highest temperature reached? *(1)*
(b) If the data logger can store 100 readings, how long could you run the data collection without overloading the memory? (2)
(c) There are two patterns in the graph. One is due to environmental conditions outside the bag. The other comes from changes inside the bag. Sketch them separately and point out possible reasons for their shapes. (5)
(d) Give two reasons for the fall in temperature at the end of two weeks. (2)
(e) The rotting process has several stages. First, the plant cells must be broken so that fungi can get inside to feed on the cell contents. Then bacteria follow and rapidly act on the partly decayed cells. Finally the remainder forms compost.
Examine the graph you sketched for conditions inside the bag and link it to these points:
(i) fungi enter plant cells, environmental warmth aids fungal growth (2)
(ii) bacteria enter plant cells, rapidly metabolise cell contents, plus some dead fungi, much energy is released (2)
(iii) most decay has finished, compost cools down. *(1)*
(f) (i) What would have first broken the plant cells? *(1)*
(ii) Where would the rot causing fungi have come from? *(1)*
(iii) How long would it be before the weed seeds were dead at 100°C? *(1)*
(g) Why did the scientist use a black bag in a sunny corner? (2)

19 Cuttle fish

Mammals have a four chambered heart. It acts as a double pump. Blood is pumped from the heart to the lungs. It picks up oxygen and returns to the heart. Then the heart pumps the blood round the body. By the time it gets back to the heart, most of the oxygen has been used up.

Ordinary blood contains haemoglobin (made up of organic compounds of iron). Haemoglobin loosely bonds to oxygen making blood bright red. De-oxygenated blood has a dark red colour. This is shown in diagram 1.

Diagram 1
Simplified diagram of blood circulation in a mammal

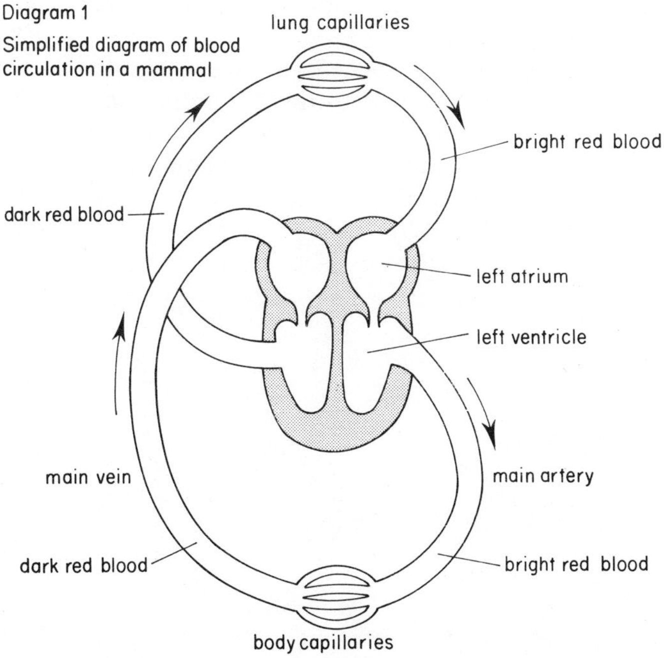

A cuttle fish has three hearts. The main heart has three chambers (two atriums and a ventricle). It has arteries which carry blood away from the main heart to all parts of the body. Branchial hearts help to pump blood through the gills where it picks up oxygen. It then returns to the main heart.
Cuttle fish blood contains haemocyanin. This compound contains copper. It is blue when it has absorbed oxygen. When it is de-oxygenated it is colourless.

Diagram 2. A cuttle fish

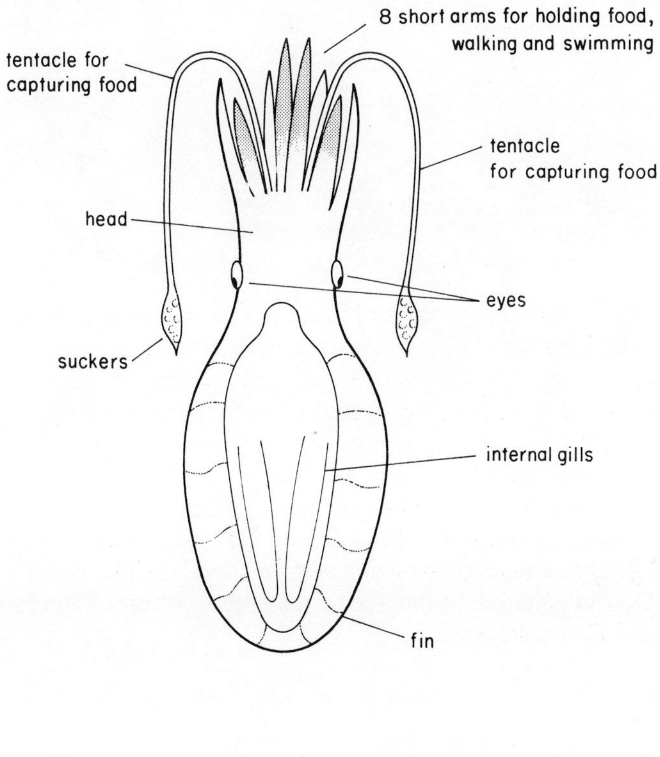

Diagram 3
Simplified diagram of blood circulation in a cuttle fish

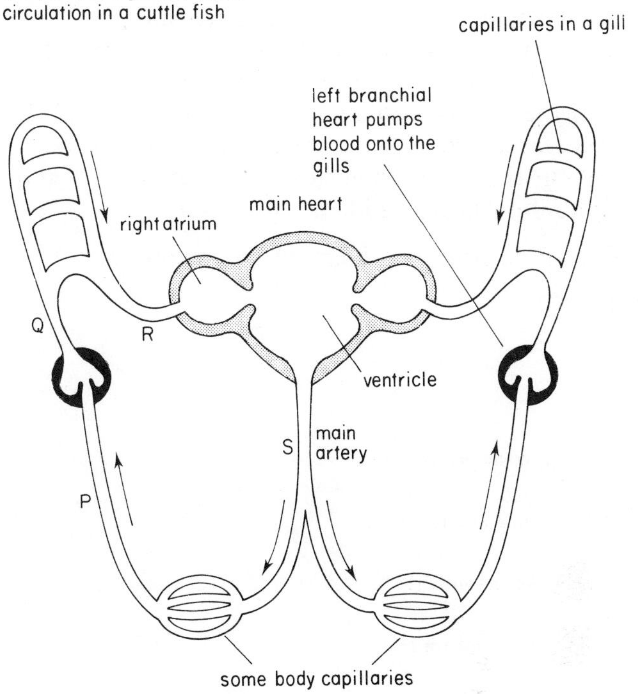

Diagram 3 shows the blood circulation.

(a) What organ does blood pass through when it goes
 (i) from P to Q? *(1)*
 (ii) from Q to R? *(1)*
(b) What is the colour of the blood at P, Q, R and S? *(4)*
(c) Where is the blood pressure highest? *(1)*
(d) Can the blood at P ever get to the left branchial heart? Explain your answer. (2)

*20 Inheritance

Certain physical features, (such as eye colour), can be inherited by children from their parents. You can show the pattern of inheritance of features in a family by using branching diagrams (Pedigree diagrams). A feature (or trait) is usually caused by only one gene, but there may be a number of different forms of this gene. These different forms are called alleles. Some alleles are dominant over others. If one parent passes a dominant allele to the offspring, this allele is the one that will show in the child's physical appearance. The alleles that are not dominant are called recessive. Recessive alleles are also passed from parents to offspring, but they are 'hidden' if they are paired with a dominant allele.

Sometimes genes are sex-linked. This means that they are carried on the X and Y chromosomes which determine the sex of the organism. As women have two X chromosomes, a recessive allele for any gene on the X chromosome can be hidden by pairing with a dominant allele. However, since men have only one X chromosome any recessive alleles on the X chromosome are never paired, and are therefore never hidden.

Use these pedigree diagrams to answer the questions.

(a) What laws of inheritance patterns are shown in these diagrams? *(8)*

(b) If ■ has to be a male with colour blindness, which is sex linked to the X chromosomes (remember male = XY), how would colour blindness be inherited by the people in the diagrams A and D? Redraw them to show what your answer is. *(4)*

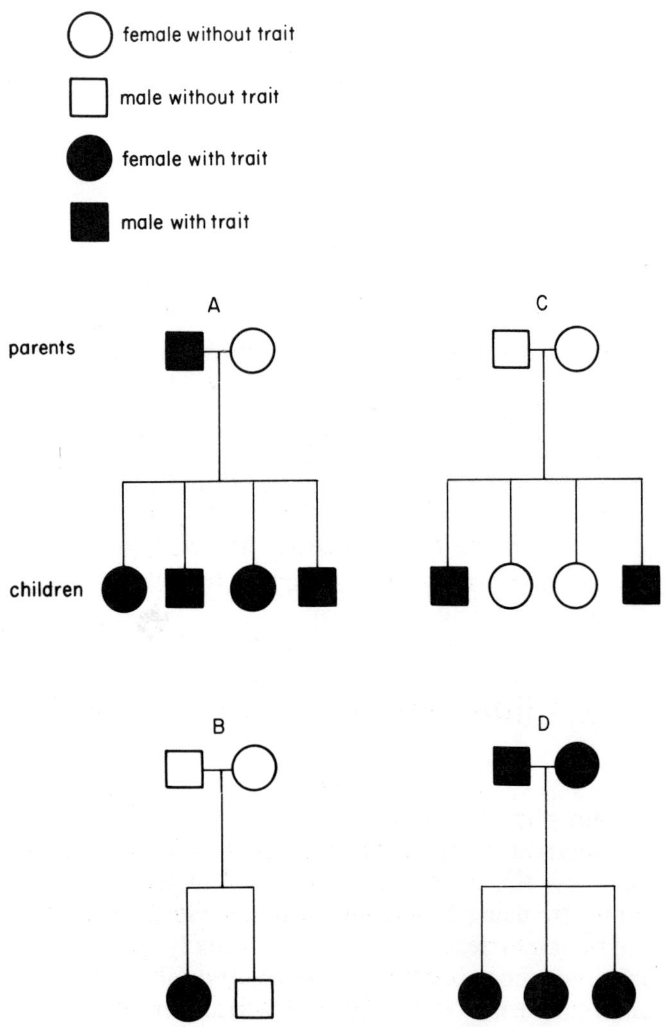

4 Evaluation

1 Egg problem

A group of pupils were asked to design and use laboratory equipment to find out the volume of the air space inside a fresh hen's egg. Below are diagrams of some of their experimental plans.

Plan 1

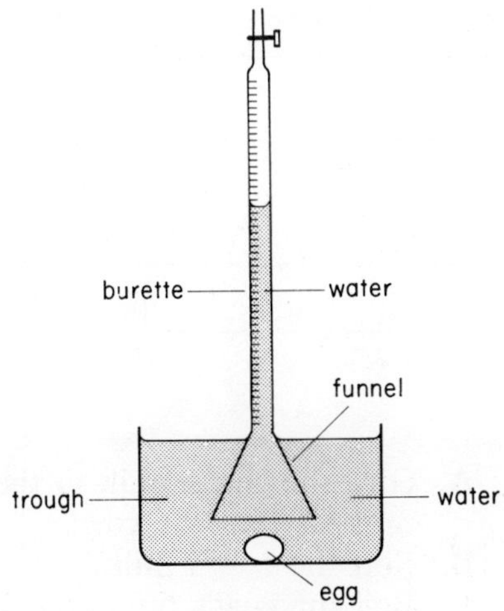

Plan 2

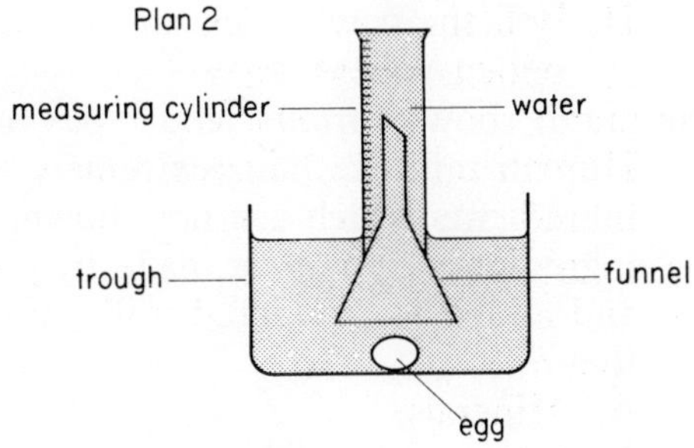

Plan 3

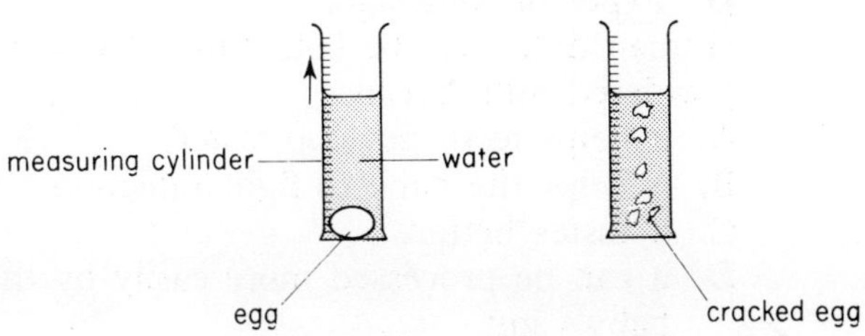

Plan 1: Crack an egg under water → the air bubble from the egg rises → the water level in the burette falls → measure volume in cm^3.

Plan 2: Similar to plan 1 but a measuring cylinder (100 cm^3) rather than a burette is inverted over the funnel.

Plan 3: Volume of egg is measured by displacement in a measuring cylinder (100 cm^3) → recorded. Egg cracked → air escapes → water level drops.

Look at the experimental plans and answer the questions below.

(a) What are the two main problems with plan 3?
- **A.** A hen's egg is too large to go into a 100 cm^3 measuring cylinder.
- **B.** The scale on the measuring cylinder is not graduated enough to give sufficient accuracy.
- **C.** The calculation of the final volume of the air space would be difficult.
- **D.** Making sure all the air in the egg was released.

(b) Which plan is likely to be most accurate?
- **A.** 1.
- **B.** 2.
- **C.** 3.
- **D.** 1 and 2 are equally accurate and better than 3.

(c) What is the main advantage of plan 3 over the other two?
- **A.** You do not have to make sure all the air is released.
- **B.** It does not matter if the air escapes into the atmosphere.
- **C.** No calculations are necessary.
- **D.** The egg is not broken during the experiment.

2 Mother's milk

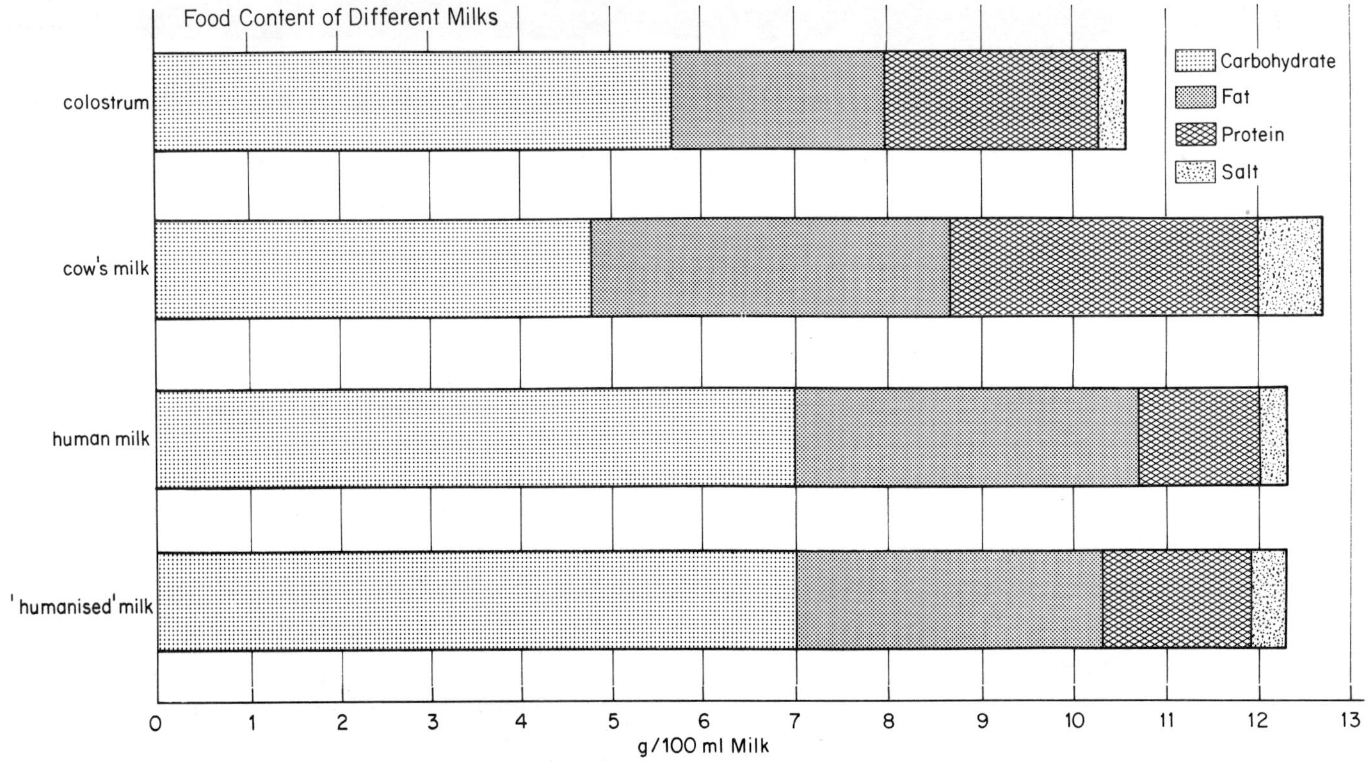

Colostrum is the milk produced by a woman for the first few days after giving birth. 'Humanised' milk is the preparation of a modified cow's milk so that the mixture is more like average human milk.

(a) In making 'humanised' milk, scientists have been fairly successful in matching some food types. Which food type has exactly the same proportion in human and 'humanised' milk?
A. Carbohydrate.
B. Fat.
C. Protein.
D. Salt.

(b) (i) The amount of salt in human milk is
A. 0.3 g/100 ml.
B. 0.3 g/1000 ml.
C. 0.4 g/100 ml.
D. 0.4 g/1000 ml.

(ii) Suppose you are stranded on an island with a three week old baby. On the island, cattle can be milked. What should you do to the cow's milk before you feed it to the baby?
A. Give the cow's milk to the baby as it is milked from the cow.
B. Boil the cow's milk.
C. Boil the cow's milk and dilute with water.
D. Boil the cow's milk and dilute with boiled water.

(c) The graph shows various food types in milk.
(i) Human milk contains extremely important ingredients which are not shown. These ingredients cannot be made by scientists and added to powdered milk. What are they?
A. Minerals.
B. Immunoglobulins.
C. Sugars.
D. Types of roughage.

(ii) Human milk may be better for babies than powdered milk because
A. it helps teeth develop faster.
B. it helps the baby to fight infection.
C. it tastes better.
D. it can be processed more easily by the baby's gut.

3 Worms

On a school fieldcourse, one group of pupils took a 50 cm × 50 cm frame (or quadrat) into a woodland, while another group took a similar quadrat to an open grassland.

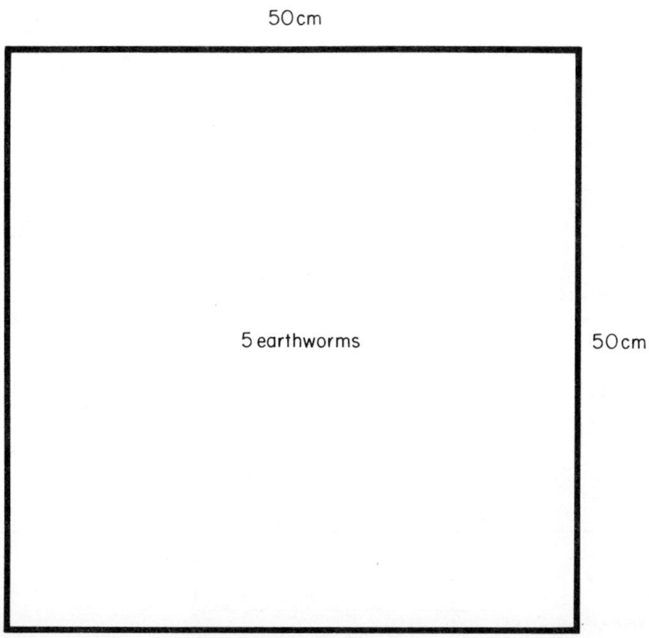

formaldehyde solution

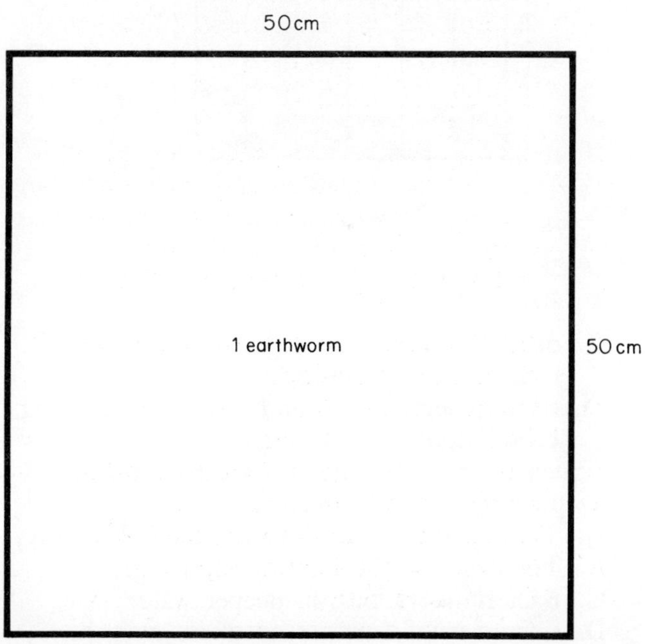

detergent solution

The group in the woodland dropped the quadrat randomly onto the ground and then poured 2 litres of formaldehyde solution on the area marked inside the quadrat. The other group followed the same procedure in the open grassland using 2 litres of detergent solution. Both groups waited ten minutes and then counted the number of earthworms that came up due to the solutions irritating them. The results are shown to the left.

(a) How many earthworms seem to be present per hectare (10 000 m^2) in the woodland?
- **A.** 50 000.
- **B.** 100 000.
- **C.** 200 000.
- **D.** 250 000.

(b) The two groups compared results and one person concluded that there are more earthworms in one hectare of ground in the woodland than there are in one hectare of ground in open grassland.
This conclusion is not valid because
- **A.** when calculated per hectare, there are more earthworms in the ground in the open grassland.
- **B.** the areas measured were different in each case.
- **C.** the amount of solution used by each group was different.
- **D.** the type of solution used by each group was different.

*4 Lake pollution

The map overleaf shows a lake and the area surrounding it.

You are a scientist employed by the owner of the lake to increase the catches of fish and to stop pollution of this lake.

(a) Your pollution problem seems to be
- **A.** acid rain.
- **B.** nitrate pollution.
- **C.** sewage pollution.
- **D.** radiation pollution.

(b) The best solution to the problem would be to
- **A.** seal up the land field drains.
- **B.** divert the stream away from the lake.
- **C.** stop the farmer using nitrogenous fertilizers.
- **D.** place some kind of filter in the stream.

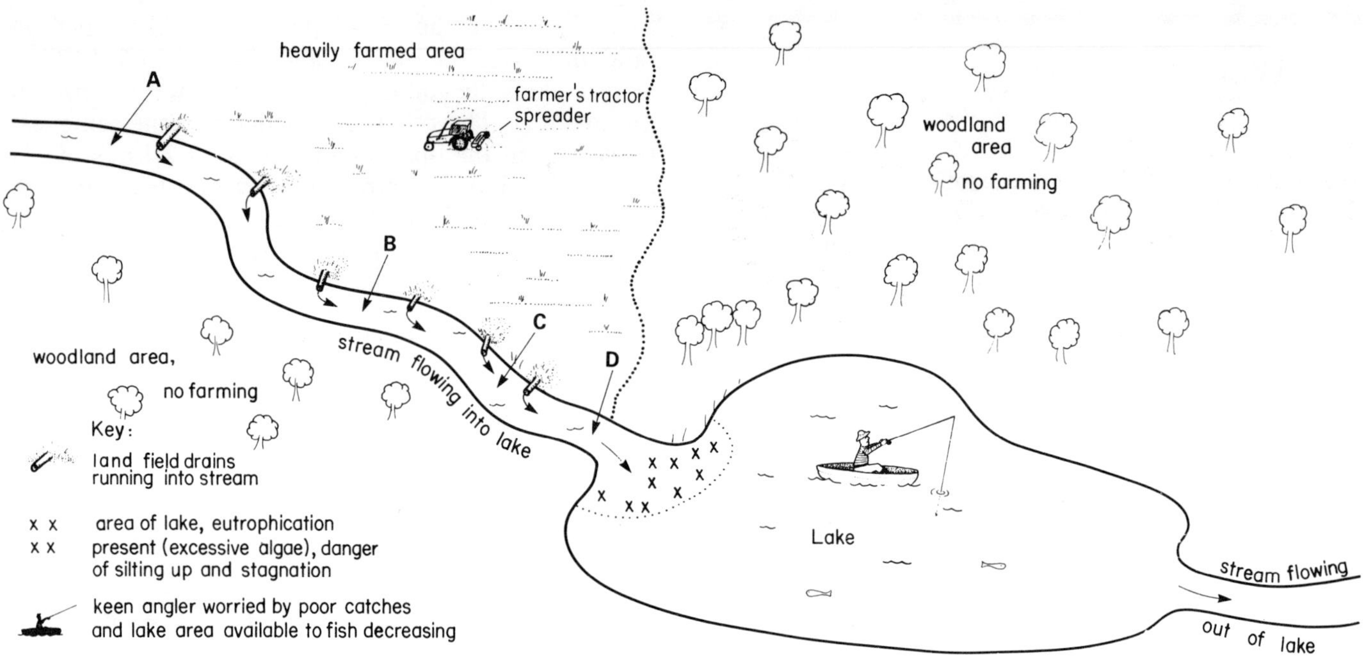

(c) A filter placed in the stream would work best at site
A. A.
B. B.
C. C.
D. D.

(d) The accompanying diagrams show a type of filter which could be used.
The filter works because
A. nitrates are trapped by the straw.
B. straw neutralizes nitrates.

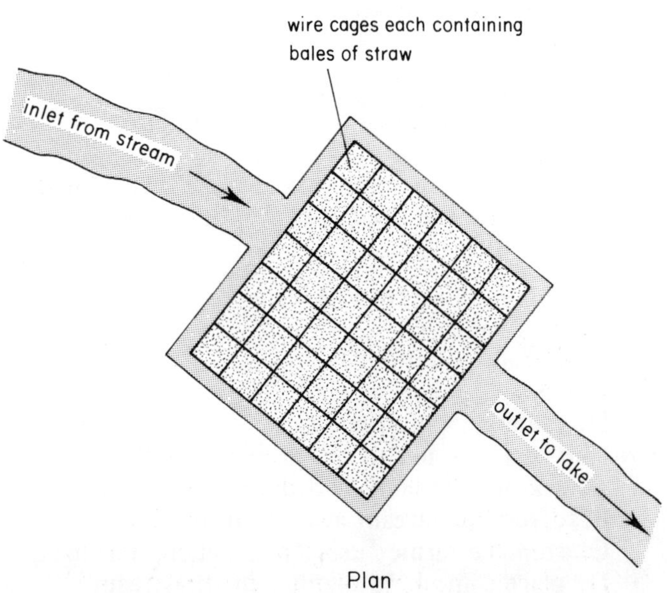

Plan

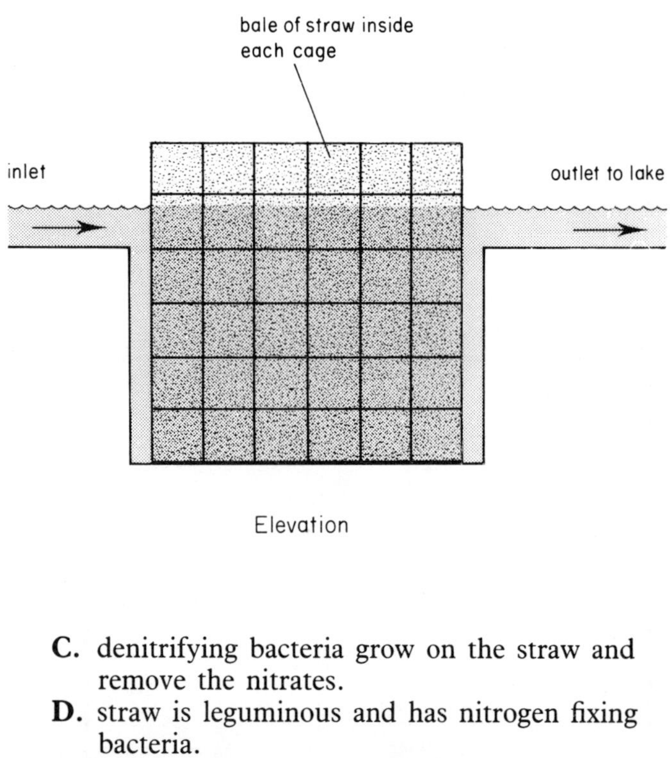

Elevation

C. denitrifying bacteria grow on the straw and remove the nitrates.
D. straw is leguminous and has nitrogen fixing bacteria.

(e) Which of the following is a good reason for keeping the straw bales in wire cages?
A. To stop the stream washing the bales away.
B. The cages stop the straw silting up.
C. Bacteria work best in deeper water.
D. Corrosion of the wire produces oxygen which the bacteria need to survive.

5 Fish

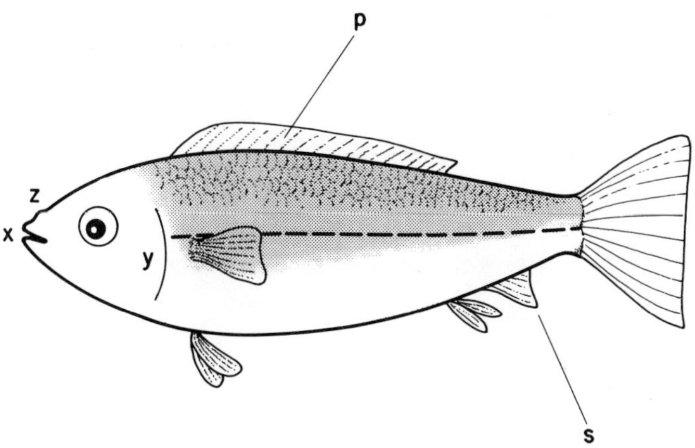

(a) This fish is darker on top than underneath so that
  A. it can blend in with the weeds.
  B. it cannot easily be seen from above.
  C. it is protected from strong sunlight.
  D. it will attract the opposite sex.

(b) The fins on its back P and underneath S are for
  A. changing direction.
  B. swimming along.
  C. keeping upright.
  D. keeping level.

(c) This fish can see
  A. only in front.
  B. only to the sides.
  C. all around itself.
  D. nearly all round itself.

(d) This fish breathes by taking water
  A. in at X and out at Y.
  B. in at Y and out at Z.
  C. in at Z and out at Y.
  D. in at Y and out at X.

(e) The fish has a line of nerve endings along its side. They sense pressure change. This helps the fish to detect
  A. bright light.
  B. harmful chemicals.
  C. warm and cold water.
  D. water movements.

6 Schistosomiasis

Schistosomiasis (also known as bilharziasis) is a group of diseases affecting the urino-genital and gastro-intestinal systems of humans. It is caused by worms (trematodes or flukes) of the genus *Schistosoma*. This is one of the most common parasitic infections in the tropics. *Schistosoma* have interesting life cycles, summarised below:

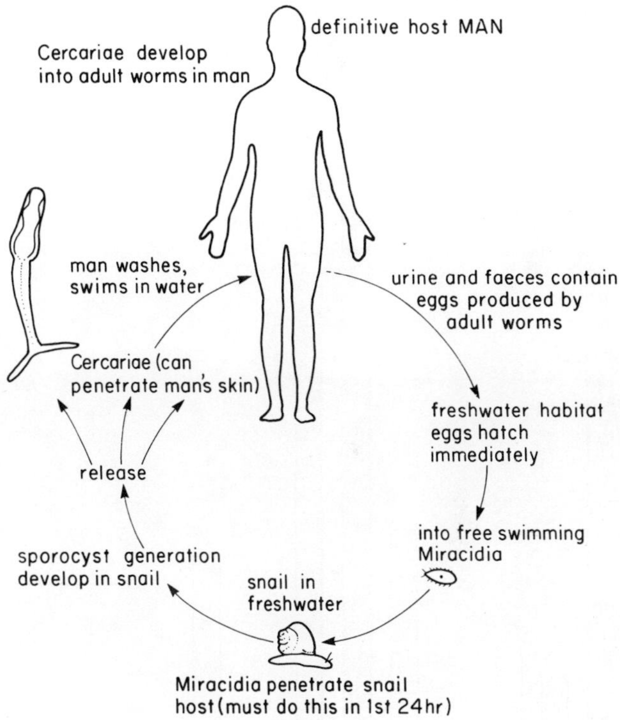

Which is the best solution to this disease?

A. Protection of freshwater habitats by provision of clean water and toilet and sewage treatment facilities.
B. Snail destruction by use of molluscicides.
C. Mass treatment of people by drugs that kill the adult worm.
D. Avoid infection by wearing protective clothing.

7

Some pupils carried out an experiment in summer to find out which colour insects preferred. Four coloured trays were laid out in a line. Each one had the same amount of soapy liquid. The trays were collected at 10.00 am and replaced with fresh ones. This was repeated at 2.00 pm and 6.00 pm. The results are plotted on bar graphs below.

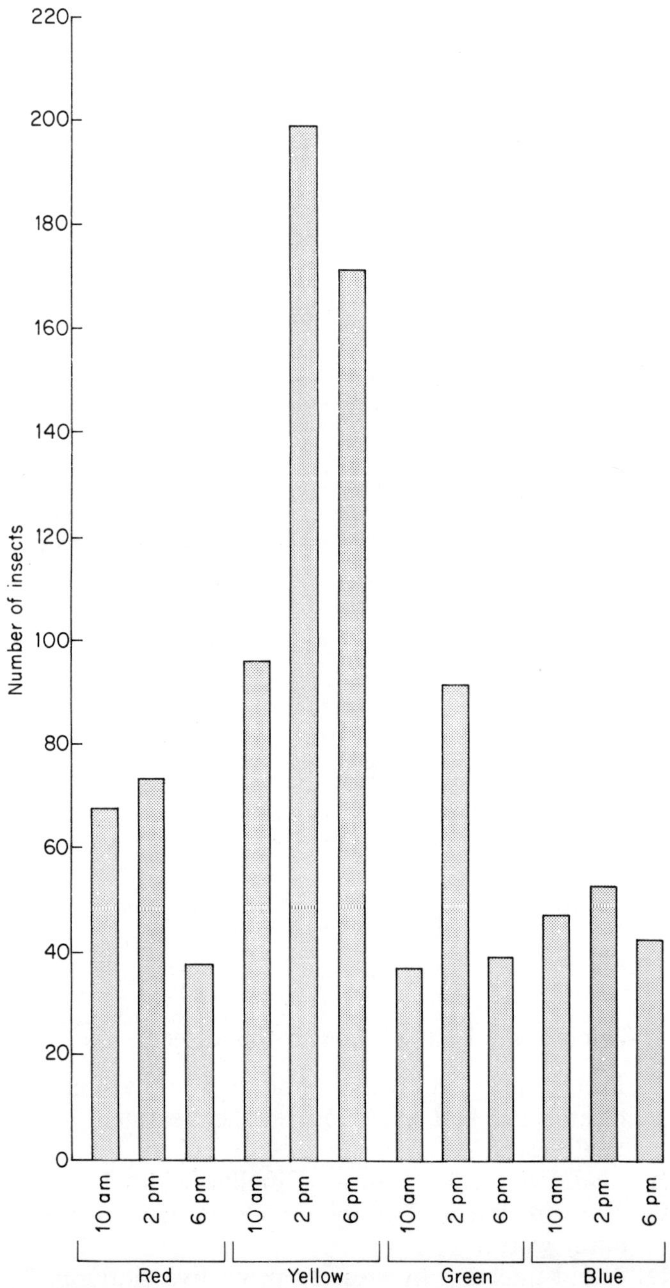

(a) At what time of day were most insects collected and how many in total? *(1)*
(b) Suggest an explanation for this. *(1)*
(c) When were least insects collected? *(1)*
(d) Which tray did the insects seem to prefer? *(1)*
(e) Why was soapy water used and not just water? *(1)*
(f) You cannot really draw firm conclusions from your answer to (d) because the trays were in a line and not in the same position. What ought to have been done to remove this possible source of error? *(1)*

*8

All organisms need to be adapted (suited) to their habitat. Three adaptations of Water Milfoil (a pondweed) are listed below. Try to think how they help the plant's survival.

Adaptation	*Survival value*
Long floppy stem	
Leaves only present at the surface of the water	
Flowers found above the surface of the pond	

*9

Try to explain the following observations:

(a) Some pond animals can rest on the surface of a pond. *(2)*
(b) Chalky streams have more snails than softwater streams. *(1)*
(c) Ponds can freeze over more easily than streams. *(1)*
(d) Some organisms are found only in certain parts of a stream. *(1)*

10 Crop rotation

If farmers plant the same crop in a field year after year, the soil becomes deficient (low) in certain minerals. Some farmers add fertilizers to prevent this. Others rotate crops, planting different ones each year. An example is shown below:

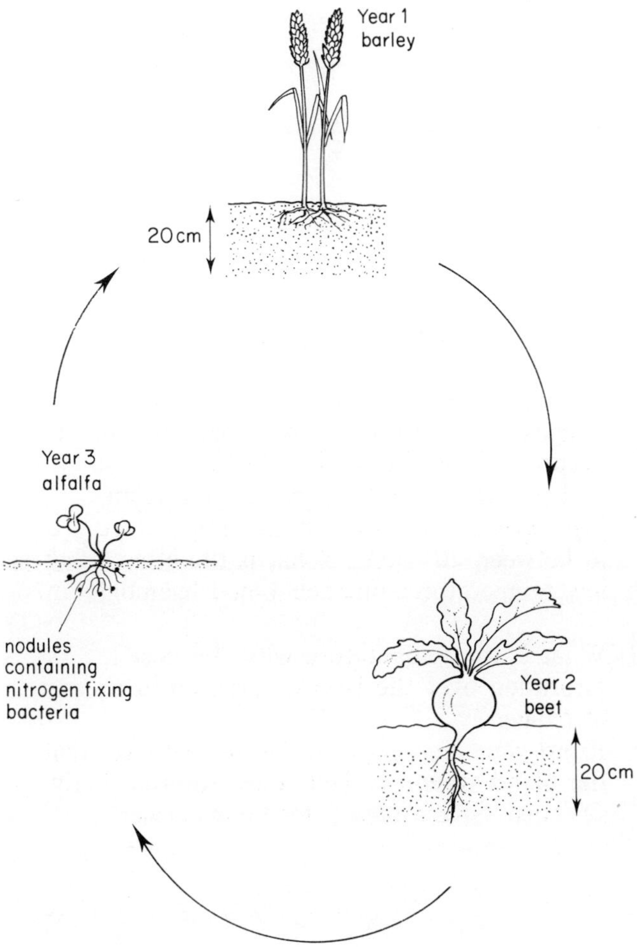

(a) From what section of the soil do barley plants get their minerals? (*1*)

(b) Suggest a reason why the farmer might plant beet next year. (*1*)

(c) Which of the crops shown opposite could be grown instead of beet? Explain your answer. (*2*)

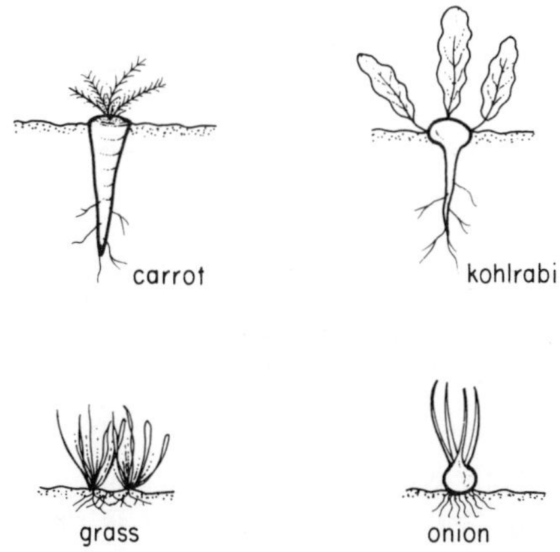

(d) Alfalfa has nitrogen-fixing bacteria attached to its roots. These bacteria can change nitrogen gas in the soil air into nitrate minerals in the soil. What should the farmer do at the end of year 3 before planting barley? Explain your answer. (*2*)

(e) Why are the nodules of alfalfa fairly close to the surface of the soil and not deep down? (*1*)

11 Camels

Camels live in hot, dry, windy climates, such as the deserts of the Sahara. Temperatures in a desert vary from 40°C during the day down to 5°C at night. The days are almost cloudless. Look at the diagram.

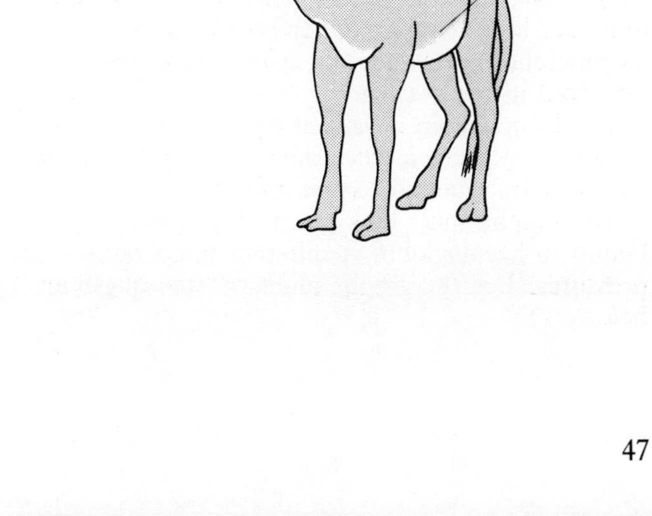

(a) How do the features labelled on the diagram help camels in these conditions?
  (i) Nostrils it can close. (*1*)
  (ii) Hairy upper surface, few hairs on belly. (*2*)
  (iii) Tough lips. (*1*)

(b) Suggest another way in which camels are adapted (suited) to life in desert regions. (*1*)

*12 Red blood cells

Red blood cells contain a pigment called haemoglobin. Most of this pigment is protein. The protein is arranged in four units and each unit can carry an oxygen molecule.

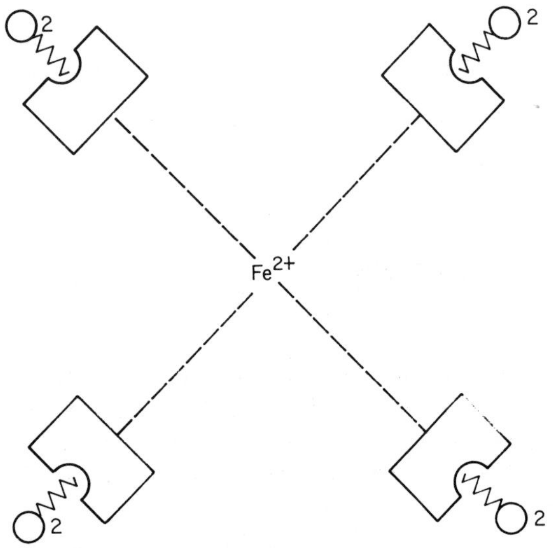

Oxygen becomes attached to haemoglobin units when it is dissolved to a high concentration in the blood plasma. The concentration of dissolved oxygen is measured in pressure units, so when the pressure units of oxygen in plasma are high, haemoglobin becomes saturated by oxygen. The oxygen *binds* to the haemoglobin. When pressure units are low, oxygen already bound to the haemoglobin is released, or *unbinds* to become dissolved in the plasma.
N.B. Haemoglobin is said to be 100 per cent saturated by oxygen when all the spaces for oxygen are filled by a bound oxygen molecule.

The graphs show how much oxygen becomes bound to haemoglobin at different temperatures and pressures. Use the graphs to answer the questions below.

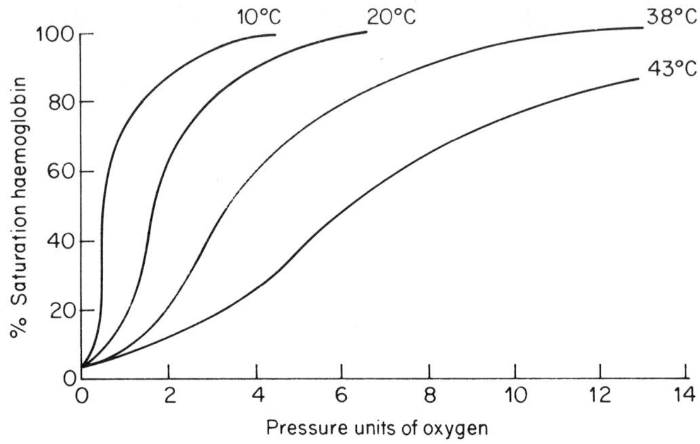

(a) How does changing temperature affect the percentage saturation of haemoglobin? (*1*)

(b) Remember that haemoglobin is a protein. Suggest one reason why haemoglobin cannot become 100% saturated at 43°C. (*1*)

(c) Old people are said to be suffering from *hypothermia* when their body temperature drops to between 20°–30°C. What is the effect of these low temperatures on their blood haemoglobin? (*1*)

(d) What is the temperature with the best saturation over the largest variation in pressure of oxygen? (*1*)

(e) From your answer to (d), would you say that the blood system is well tuned to other body systems? Give a reason for your answer. (*3*)

*13 Bracken

Bracken is a poisonous fern which contains many toxins and carcinogens. It is thought to cause human stomach cancer through milk, meat and water by contaminating pasture. In rural areas where there is a lot of bracken, there is often an unusually high death rate due to stomach cancer. Some towns which obtain their water from moorland areas with bracken have a similar problem.

Bracken grows well only in full sunlight. It does not grow in the damp frost hollows of valley bottoms nor on windy hilltops. It is found on the middle slopes where shelter combines with fairly good soils.

(a) (i) Explain why farmers do not want bracken. (*2*)
(ii) Why does it have no natural enemies? (*2*)

(b) There is a high death rate from stomach cancer in the villages of Farmham and Sheepham. There is little stomach cancer reported in Burton, Busk, Greenwood and Lavington. Copy the map and show on it areas with a lot of bracken. Also show where a river could be dammed to make a reservoir for Muckton. (*4*)

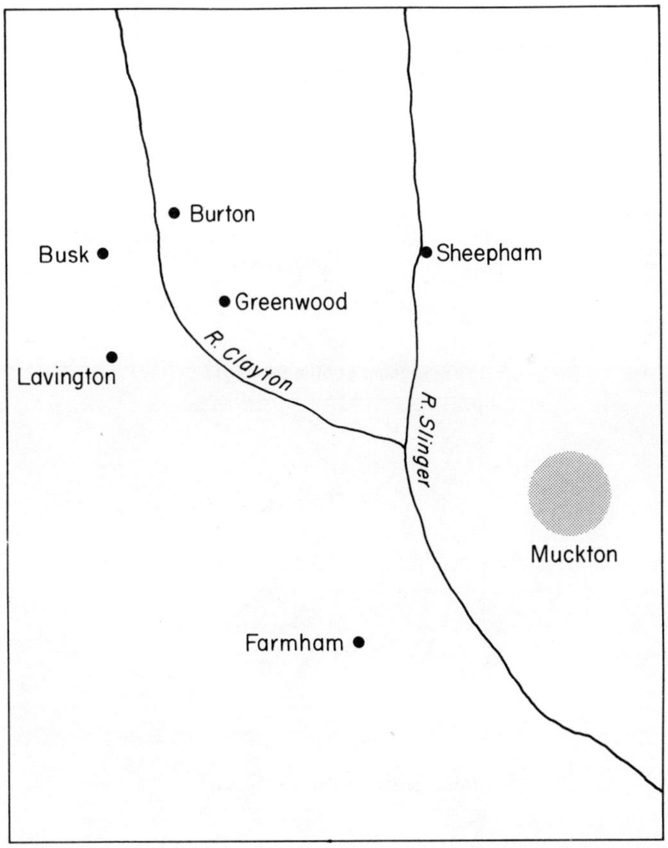

(c) There is a strong argument for planting bracken infested hillsides with coniferous trees. Explain the reason for this. (*4*)

(References: Coleman, A. The place of Forestry in a viable land-use strategy. *Quarterly Journal of Forestry LXXIV* 1980
Howe, M. *National Atlas of Disease Mortality in the United Kingdom* 1970)

14 Cats

(a) Name three things which tell us that cat 1 is a hunter. (*3*)

(b) Cat 2 is nearly asleep. Explain why it is stretched out, long and thin. (*2*)

(c) Cat 3 is in a cold place. Why is it snugly curled up? (*1*)

(d) Could there be a temperature difference between the parts marked A and B on cat 3? Explain your answer. (*2*)

(e) Cat 1 and cat 2 have bristly whiskers. One is asleep and one is awake. How does each rely on its whiskers? (*2*)

*15 Getting abreast of a problem

In a survey, many women were anxious that breast feeding or not breast feeding their babies may alter the shape of their breasts.

This is how one scientist went about collecting data and evidence and evaluating them. The scientist came up with the following plan. Criteria:

1. Size of breasts measured by chest diameter (cm)
2. The degree of contour measured by the angle the upper surface of the breast made to the chest wall (degrees °)

Both these measurements were taken before pregnancy and two years after pregnancy. The women studied were put into two groups, those who had breast fed their babies and those who had not.

The degree of contour was to be assessed accurately by photographing the breast outline from the side view before and after pregnancy in both groups of women. The angles of breast to chest were compared from the photographs. The results are shown in the diagrams.

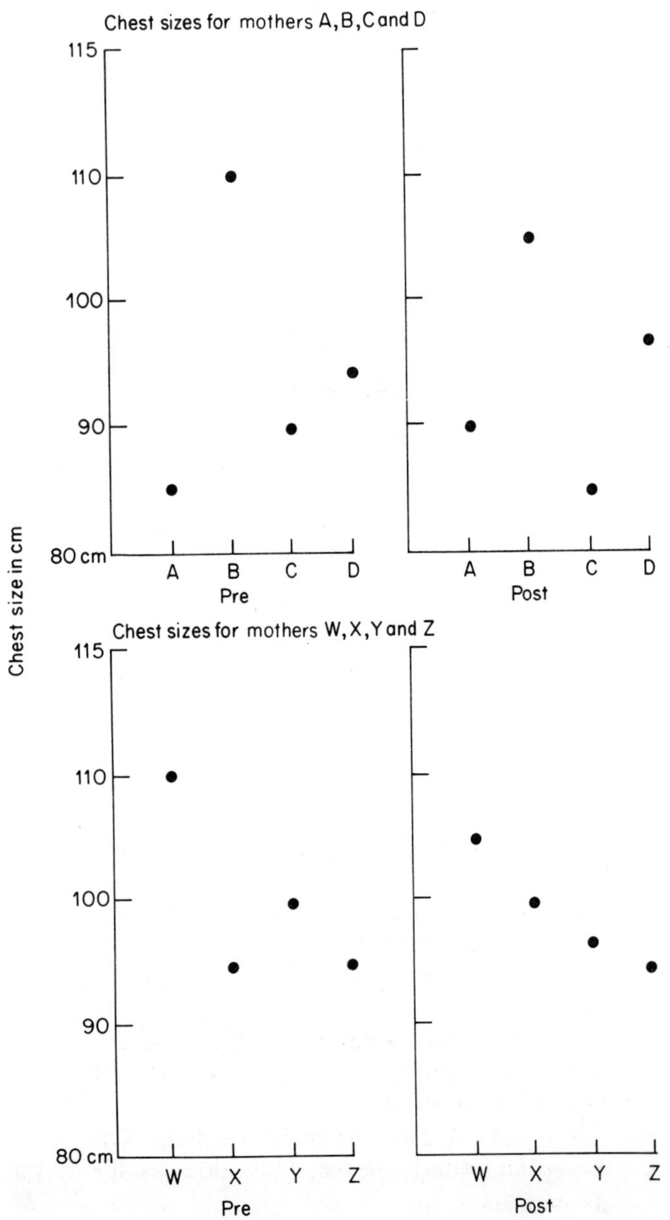

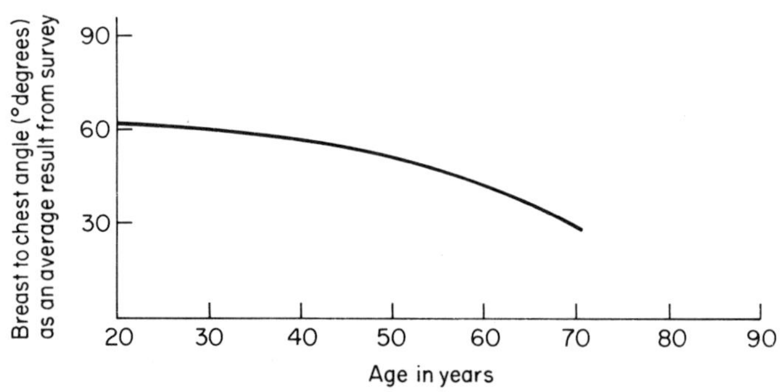

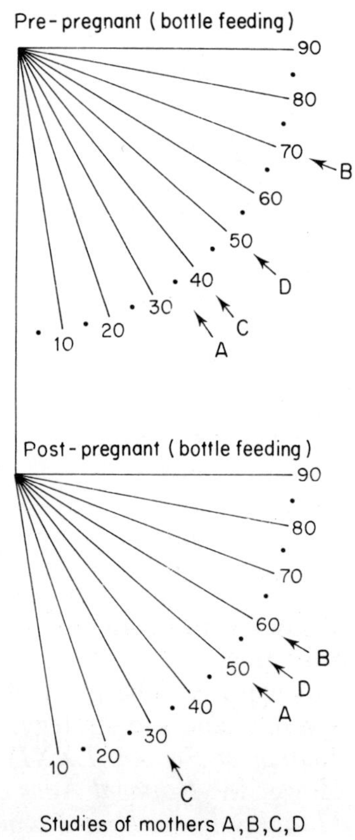

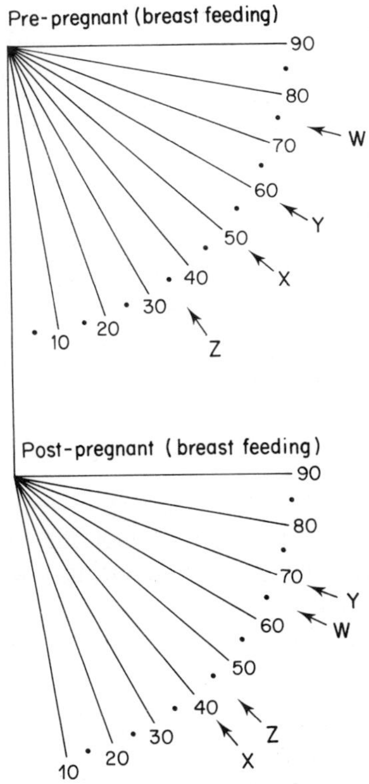

Studies of mothers W, X, Y, Z

(a) Can you think of any criticisms of this scientist's methods so far? Where may errors be present? *(5)*

(b) What are your conclusions from the data? Does pregnancy alter the breast shape? Does age alter the breast shape? Is there definite proof that breast feeding alters the breast contours more than bottle feeding? *(3)*

(c) Which of the mothers had the greatest overall change to her breast contours?
A. Mother A.
B. Mother B.
C. Mother X.
D. Mother Y.

(d) Do you think this was a worthwhile scientific survey? Do you think the money could have been better spent on other kinds of research? Give your reasons. *(4)*

(e) (i) After what age did the survey show the greatest rate of fall in breast to chest angle?
A. 20–30 years.
B. 30–40 years.
C. 50–60 years.
D. 60–70 years. *(1)*

(ii) Why do you think such a change in a woman's body happens at this time of life? *(1)*

A

*16 Advice on weed control

Read the information given below about common crop weeds. Then evaluate the advice given to the farmer as he asks questions.

Weeds growing in crops are a nuisance. This is because they compete with the crop plants like oats and barley. They take up space, soil nutrients and light. Weed biologists have made some discoveries about weed plants. These discoveries explain the success of certain weeds and help us to control them.

Weeds usually have a high percentage germination. This means that about 80 per cent or 90 per cent of weed seeds will germinate. Fewer of the crop seeds (with which the weeds are growing) will germinate. The life cycle of a weed is often short, so that several generations can be fitted into the growing season. This has a crowding effect.

Weed seeds are more viable than crop seeds. This means that they are hardy and less likely to be destroyed by bad weather.

As some weeds grow, they push other plants out of the way, e.g. daisy and plantain plants flatten out their leaves to get space. Thistles grow tall and bushy and push other plants out of the way. This also shades other plants.

Crop sowing and weed germination times might coincide. Weeds germinate more quickly than crops. This gives the weeds an advantage. Weather conditions favour weed growth, but not specially-bred crop plants.

Soil quality is important. The soil pH must suit the crop but not the weeds.

Some weeds are adapted to compete with specific crops by poisoning them.

Some examples

Wild Oats are more aggressive than barley. They crowd the barley.

Camelina kills flax by secreting flax poisons from its leaves to slow flax growth.

Shepherd's Purse produces 3,700 seeds per plant per season. These seeds will germinate all year round, but especially in February, March, April, August, September and October. Shepherd's purse seeds can still germinate after sixteen years.

Black Nightshade seeds germinate in June and July. After forty years, 89 per cent will still germinate.

Blackgrass can compete with wheat on soils low in potassium.

Couch grass can take up nearly 75 per cent of its potassium needs by July.

Fat Hen can develop a full, large root system in twenty-one days.

Maize starts a major growth period at the end of June if the weather is good.

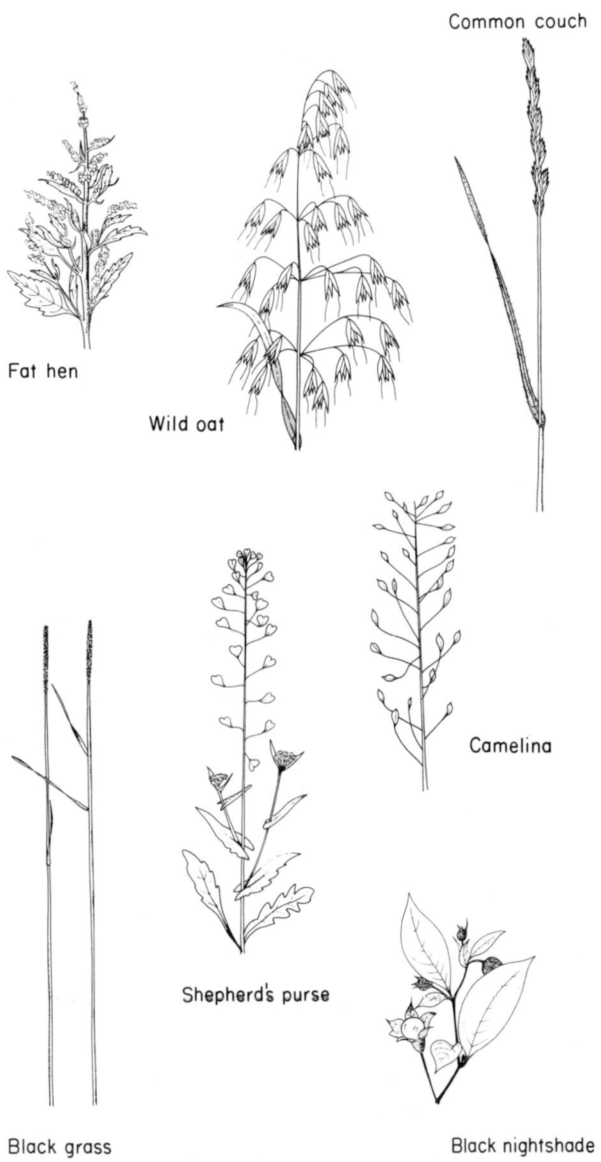

Use the information given above to evaluate the advice given to the farmer (F) by the advisor (A). Comment on statements marked **.

(a) F: I have three difficult fields. They might all give poor crops and I must find ways of improving crop yields. My south facing field with a good slope is a maize field, but this year the crop was poor. I know that the weather was too cold before June, but it was plagued by couch grass as well.

A: Well, you need to use a fertilizer for the maize. One with high K (potassium) content would help as your crop is a seed crop.** This fertilizer would also correct the effect of greedy couch grass which must have affected your maize plants.** *(4)*

(b) F: Yes, I'll try that. Perhaps I'll look for an earlier growing maize, but I must leave time to get rid of the shepherd's purse before I plant.** *(2)*

(c) A: You could, but I don't think you will get rid of it completely.** Get the timing right for adding fertilizer, weeding and sowing. What will you do first?

F: I'll have to see what new seed there is. Weed, fertilize and sow, I expect.** *(3)*

(d) F: Now what about my barley? We always have trouble with wild oats. They are in the field and there are always some contaminating the seed. They crowd out the barley seedlings.** *(2)*

(e) A: Use our patent wild oats killer. Spray it on. It won't harm anything else.** *(2)*

(f) F: How many times do I spray?

A: Twice. Once when they start to germinate and then as they grow, about two weeks later. The spray is biodegradable.** *(1)*

(g) F: I'll look into that. Now on my other field, I can't decide whether to try wheat or flax. There is a lot of couch grass and blackgrass and as you know, my land gets low in potassium.

A: In your case, you can choose what you like, but make sure that there is enough potassium or the wheat will not do well** if the blackgrass is not kept down.** *(1)*

(h) A: Flax might be a problem if you don't get good seed guaranteed free of camelina. Once you have to eliminate camelina you have a problem for years.** *(2)*

(i) F: Yes, I'll think about that. I might stay with wheat for another year. I suppose there is no way of controlling black nightshade in the hedges? Children frequently walk along the footpath.

A: No, not really. That's another one that seems to last for ever. No sooner do you destroy a plant than another grows to take its place.** *(2)*

17 Herbal remedies

Before modern drugs became available, plants were used to ease the symptoms of illness. We still use some of these remedies. Extracts from foxgloves give us the drug *digitalis*. Aspirin was first identified in extracts of willow.

Gerard was a herbalist in the seventeenth century. Here are some of his remedies.

> You will know about angelica, a green stalk that is crystallized and used for cake decorations. 'Angelica roots are a remedy against the plague, poison and infections from evil and corrupt air. Chew it between your teeth to keep away pestilential air.'
> 'Camomile is very good for all aches and pains, bruises, lumps and swellings.'

Galen, a Greek physician (129–200 AD) recommended:

> 'Paeony is very good for epilepsy. Tie the fresh root about children's necks and they will be protected.'

Other books referred to animal life as well.

> 'Take some woodlice and seethe (simmer) them in oil. A few drops in the ear will cure earache.'
> 'Wrap a spider in a ball of wet bread. Hang it around a child's neck and he will not catch whooping cough.'

By reading these remedies we can learn about theories of disease in earlier times. The questions below are about these theories.

(a) We know now that epilepsy is caused by a malfunction of the brain. Did Galen consider that epilepsy was caused by such a malfunction? Write three lines showing how his ideas about the cause and cure for epilepsy differ from ours today. *(3)*

(b) Could 'woodlice oil' have relieved earache? Give a reason for your answer. *(2)*

(c) We often rub essences and oils on the chest and throat of someone with a heavy cold. So what is wrong with hanging spiders and paeony root around children's necks? *(1)*

(d) We now know that camomile contains a drug. What kind of a drug might it be? *(1)*

(e) Gerard believed that disease was carried by pestilential air. How far is this theory likely to serve as a basis for treatment? *(3)*

18 Kidneys

Kidneys are concerned with regulating water levels and excreting excess nitrogen. The ways in which nitrogen is excreted vary. They include:

(i) Ammonia—this is extremely toxic (poisonous) but it dissolves readily in water.
(ii) Urea—this is less toxic but also soluble.
(iii) Uric acid—this is non-toxic and insoluble. It forms a paste.

Study the table below.

Animal	*Where it lives*	*Availability of fresh water*	*Loop of Henlé present*	*Main excretory product*
Mammals	land	poor	yes	urea
Birds	land	poor	v. few	uric acid
Snakes and lizards	land	poor	no	uric acid
Fish	freshwater	good	no	ammonia
Turtles	aquatic(sea)	poor	no	urea and uric acid
Frogs	land and freshwater	poor	no	urea
Tadpoles	freshwater	good	no	ammonia

(a) Evaluate the following statements:
State whether the statement is
A. totally true for all cases
B. true for most cases
C. false for most cases
D. false for all cases.

Justify your answers.

(i) Ammonia is the major excretory product of animals which live in freshwater. *(3)*
(ii) Urea is the waste product when freshwater is plentiful. *(3)*
(iii) Organisms with loops of Henlé do not make uric acid. *(3)*

(b) What is the main pattern for animals living on land or in the sea? *(2)*

*19 Synergism

Sometimes in medicine two or more drugs can be used together to have a greater effect on an illness. This is called a synergism. In a similar way factors which damage health can come together to produce a particularly bad illness.

Synergism is found in other branches of science. For example the factors which affect rusting can be tested. A simple 'analysis of variance' chart is shown below.

		Water	
		Present	*Absent*
Air	*Present*	Water saturated with air A lot of rusting	Dry air No rusting
	Absent	Boiled water A little rusting	Vacuum No rusting

(a) Similar work could be done to show the action of pepsin and hydrochloric acid on protein. Copy and complete the diagram as an analysis of variance chart for this. *(5)*

		Hydrochloric acid	
		Present	Absent
Pepsin	Present		
	Absent		

Lung damage can be caused by the gas sulphur dioxide and by smoke particles. Other things can also cause lung damage. Scientists now know that some smoke particles are small enough to settle in the lung. Some are large enough to be trapped by nasal hairs. The ones which are of middle size remain in the throat. They provide surfaces on which sulphur dioxide can be adsorbed by water to form acid. Unless experiments are well designed, this synergism can be missed.

(b) Some scientists did an experiment to try to find the cause of bronchitis and emphysema in industries where sulphur dioxide and smoke particles are present. They used laboratory animals. They decided to control variables by

1. Keeping the size of smoke particles constant (small) and varying the level of sulphur dioxide
2. Repeating 1 with large smoke particles.

(i) Draw up an analysis of variance chart for this experiment. *(3)*

(ii) Comment on the effectiveness of this experiment and suggest how it could have been improved. *(3)*

(iii) Other factors might be worth investigating to find out if they affect lung disease. Give three examples. *(3)*

(c) (i) Woodlice are sensitive to dampness, light intensity and touch. Explain briefly how the synergism between these three factors could be investigated. Show the kind of analysis of variance chart you could use. (The diagram for Question 5 in the Problem solving section might help you.) *(3)*

(ii) This is called "multivariate problem solving". It is common in biology. How would you cope with an investigation of four or more factors? Can an analysis of variance chart still be used? *(3)*

5 Synthesis

1

Animals obtain food in a variety of ways. Some of these are:

chewing solid food,
swallowing food whole,
sucking liquid food,
absorbing food through their skin,
filtering fine food particles.

Below are diagrams of several animals.

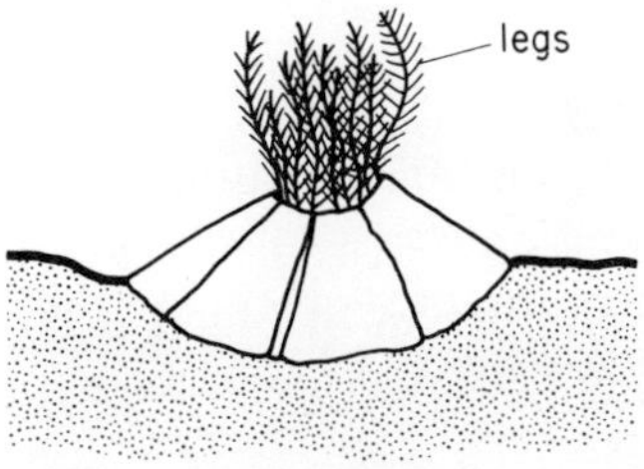

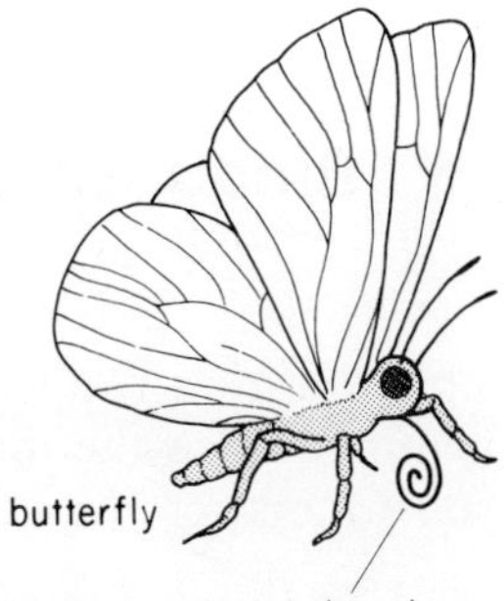

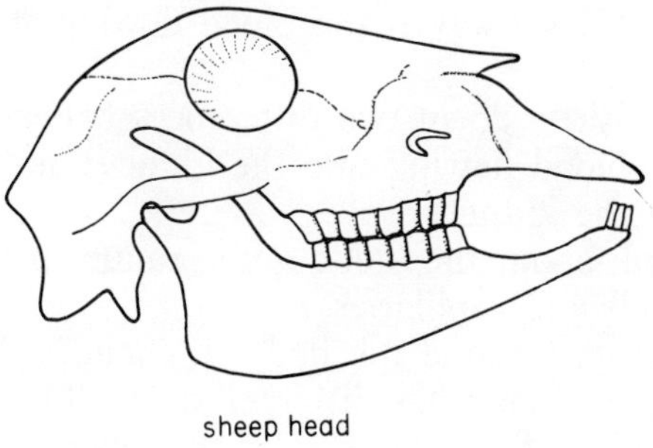

For each one:

(a) Say how it obtains its food. (Use the list to the left.) *(4)*
(b) Give one reason for each of your choices. *(4)*

2

The hammerhead shark has an excellent sense of smell. It can detect prey several hundred metres away. As it swims it swings its head from side to side.

The diagram below shows the dorsal (back) view of this shark.

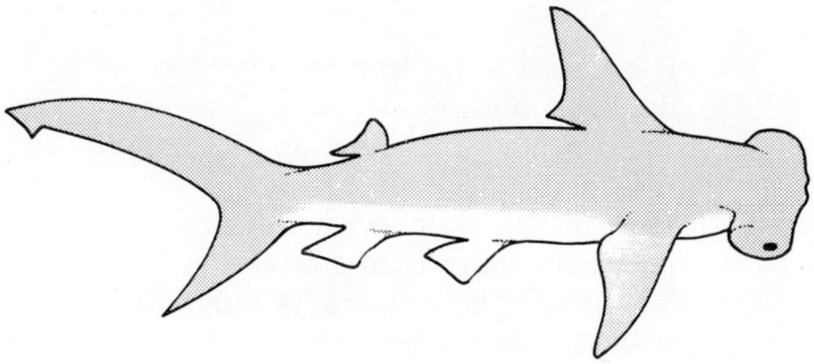

(a) Copy the diagram and mark on it where you would expect to find its nostrils. *(1)*
(b) Those sharks with bigger 'hammers' to their heads are more likely to survive than those with smaller ones. Explain this. *(1)*
(c) Suggest why these sharks need a good sense of smell despite the fact that they have eyes and ears. *(1)*

*3 Smog

(a) Warm air rises and cold air falls. Diagrams 1 and 2 show an industrial town in a valley. The weather conditions are also shown.

Diagram 1

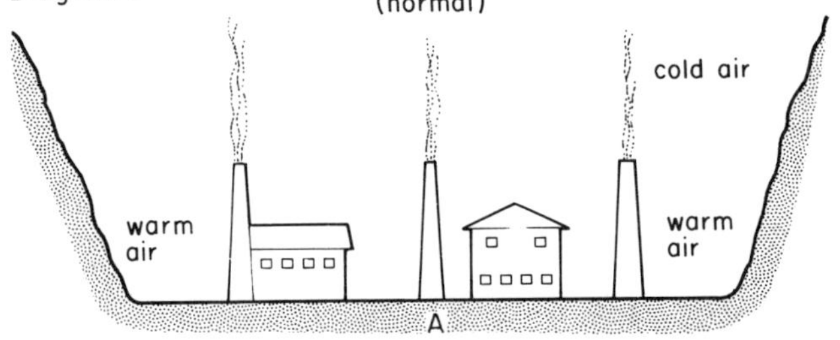

Diagram 2

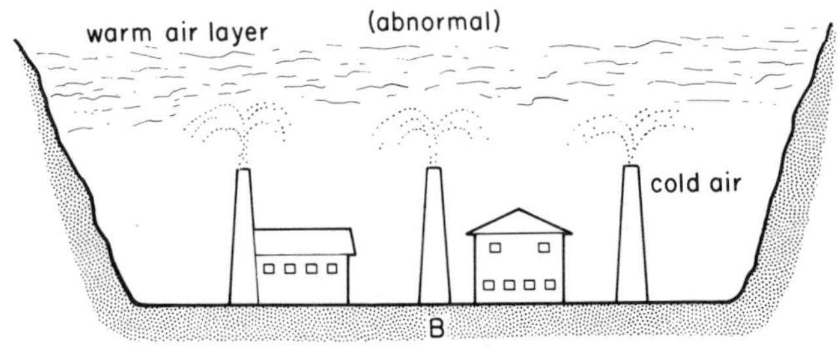

(i) B is called a temperature inversion. What does this mean? *(1)*
(ii) Explain how this inversion increases the air pollution. *(2)*
(iii) How do tall chimneys help to carry pollutants away from a city? *(1)*
(iv) Why don't these chimneys reduce air pollution on the planet? *(1)*

(b) Diagram 3 is a graph showing the effect of the London smog in December 1952.
(i) On December 5th, what was the
1. level of smoke? *(1)*
2. level of sulphur dioxide? *(1)*
3. number of deaths? *(1)*
(ii) When was the highest number of deaths? *(1)*
(iii) What is the relationship (link) between the levels of smoke and sulphur dioxide and the number of deaths? *(2)*
(iv) Name two diseases that may have led to the deaths. *(2)*
(v) What probably caused the level of smoke and sulphur dioxide in the air? *(1)*

Diagram 3

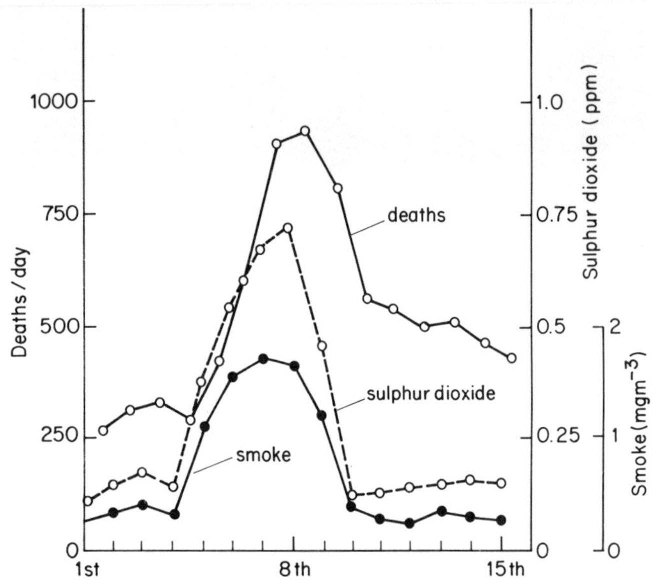

Source: *Air Pollution and Health.* A Report of the Royal College of Physicians 1970 (Pitman Medical).

*4 Blood circulation

This question is about the circulation of blood in a mammal. Use the diagram on the next page to help you to answer the questions.

(a) Use the diagram.
(i) Name organ R. *(1)*
(ii) Which way is the blood flowing at P and Q? *(2)*
(iii) Write down two differences between the blood flowing into the kidneys and out of the kidneys. *(2)*

(b) Think about the blood going round in the vessels and capillaries.
As it goes round the body, it changes. We can say that blood has different states. These are in a key on the diagram. Use the diagram to answer these questions.
(i) Does blood in state X contain more or less oxygen than blood in state Z? *(1)*
(ii) Is it true that blood in state X is never at a higher pressure? Explain your answer. *(2)*
(iii) Blood in state Y is shown only in one vessel going away from the gut. Name two substances that would be in blood going away from the gut in this vessel. *(2)*

(c) Consider the patterns in the blood circulation diagram.

(i) Why is this called a double circulation? (*3*)

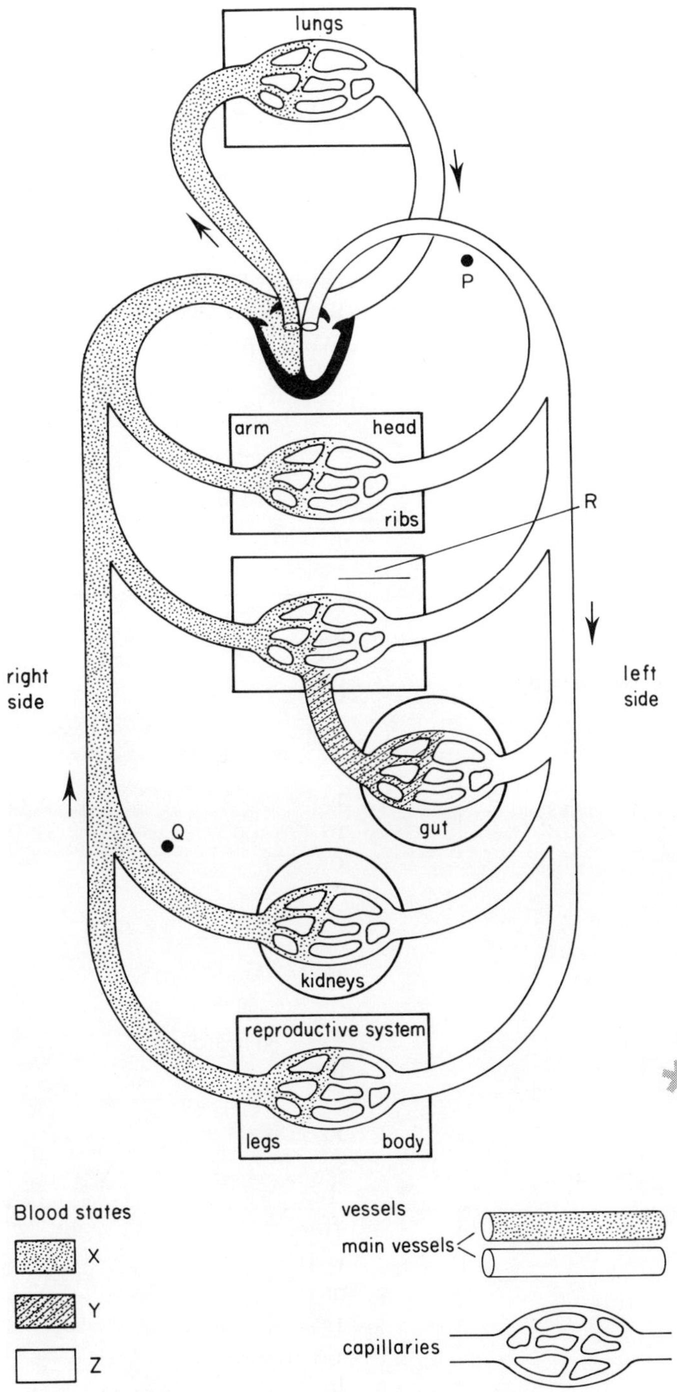

(ii) Find three differences between this diagram of the blood circulation and the real blood circulation in a mammal. (*3*)

(iii) Early medical thinkers like Galen (AD 129–200 approx) thought that the heart sent the blood out of the body from the left ventricle. Then the right ventricle sent vapours out along the pulmonary artery and the blood crossed into the left ventricle through invisible pores. Copy and label the diagram below to show what Galen thought. Mark with a tick ideas which we think are correct. Mark with a cross the ideas that we think are wrong. (*10*)

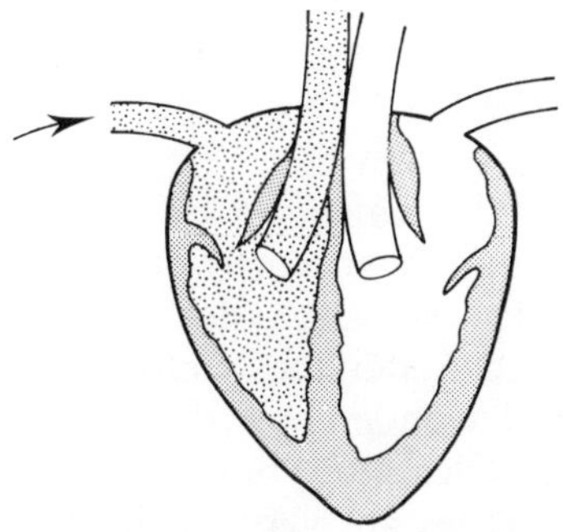

(iv) Galen's ideas were believed for many years, partly because studying human anatomy was forbidden and partly because working with living bodies (vivisection) was absolutely forbidden. Quite a lot of early medicine was worked out on animals. Give an explanation for Galen's misunderstanding of blood circulation through the heart. (2)

(v) Ibn-an-Nafis (1205–88) was an Islamic doctor. He said that blood went from the right side of the heart to the left by going through the lungs. Was he correct? Why did no-one in Europe accept his ideas? (2)

*5 Watering plants

Jane and Tracey are in the greenhouse, talking about their plants.

Jane: This one is dying.

Tracey: No it isn't. It is still green. It's just wilting. Go on, water it.

Jane: If I pour water on its leaves and the soil, then it will be OK.

Tracey: Try it. I think it would be better to water the soil. You might drown it. Crumbs, it's hot in here!

Jane: My mum puts sugar in with cut flowers if they wilt. It perks them up. Shall we try it with these plants?

Tracey: No. They can't eat sugar.
Jane: Well it works on cut flowers. They always use it in my mum's flower shop.
Tracey: Look, it's soaked up the water already. The sun's done that. Pot plants can't take up sugar. They haven't got big enough holes in their roots.
Jane: What?
Tracey: I think that only water can get into roots. Your flowers must be able to use sugar because it can go into their cut stems.
Jane: Well, there are holes in leaves. I said, "water the leaves". Maybe we could give them sugar water.
Tracey: No. The leaf holes are for gases to go in and out. You would drown it.
Jane: I didn't see any holes in the roots when I looked under the microscope. Did you?
Tracey: No. I wonder how the water gets in.
Jane: The holes must be very small. Water molecules are very small. The roots soak up the water and then it goes up the stem in the water vessels, right to the leaves.

(a) Would plants usually rely on taking up sugar from the soil? Give a reason for your answer. *(2)*
(b) Why would Jane's idea of watering plants with sugar be an idea that would not work? *(2)*
(c) What did Tracey mean when she said that the sun had helped the plants to soak up the water? Copy the diagram and show where the water would go. *(3)*

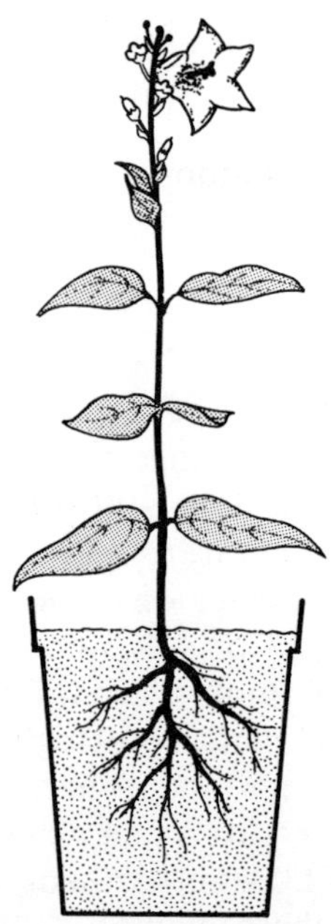

(d) Is Tracey's explanation of why sugar solution works for cut flowers but not for pot plants reasonable? Explain your answer. *(2)*
(e) Design a test to find out whether sugar improves *cut* flowers that are wilting. Explain how your test would work and what you think the answer might be. *(6)*

6

Soil scientists classify soil into various layers. The diagrams below show the layers in two soils from different areas. They are called profiles.

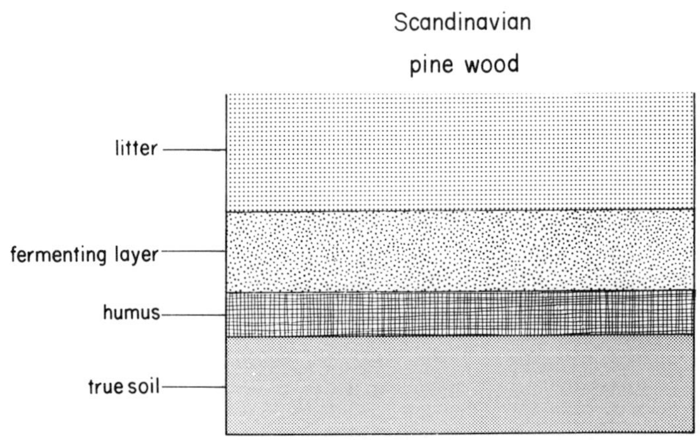

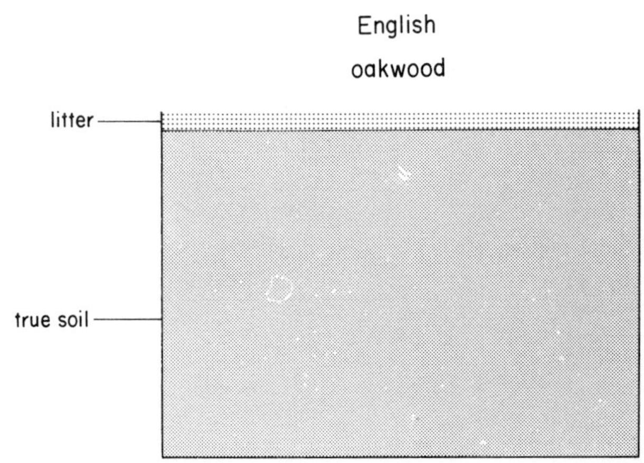

(a) Note two similarities. *(2)*
(b) Note two differences. *(2)*
The bar chart gives some data for the two areas.

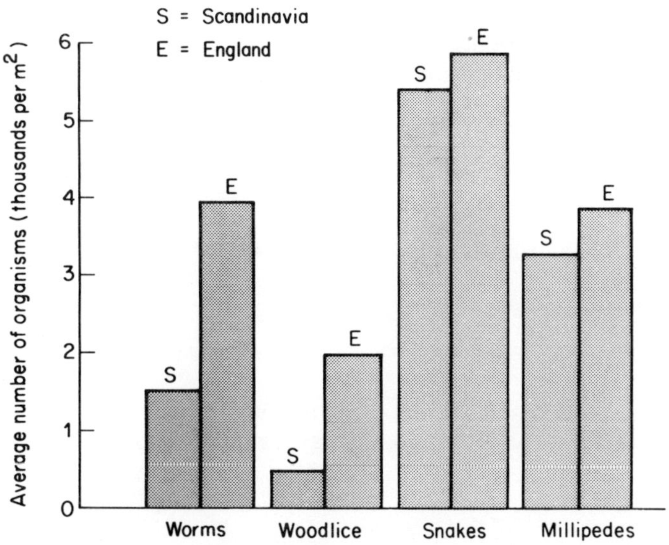

(c) Describe three pieces of information that the graph tells you. *(3)*
(d) Using the bar chart, suggest the most likely explanation for the differences that you noticed in the soil profiles. *(2)*
(e) Suggest three other explanations for the differences. *(3)*
(f) Describe clearly how you would test one of your hypotheses (explanations). Include what you do, what you keep the same, what you measure and how you avoid fluke results. *(6)*

*7

Mammals have certain features in common. One of these is that they keep their body temperatures almost constant. Even closely-related mammals living in different climates can appear quite unlike one another.

Look closely at the following diagrams.

(a) One of these animals lives in a hot climate. Say which one. List three features of its design—explaining how each feature helps its adaptation (suitability) for a hot area. *(4)*
(b) Suggest a reason why very small birds are not found in extremely cold climates. *(3)*

8 Pond

This question is about a food web in a pond. All of the information you need is in the question, but you can look up more if you think of some more questions.

Copy the pond diagram overleaf, taking about half a page. Draw the animals in place, where they live in or near the pond, then join them with red lines to show their feeding relationships. Remember that the arrows in food webs show which way the energy is going.

(a) Kingfishers eat minnows.
(b) Frogs eat fly larvae from the water surface, the banks and water.
(c) Otters eat larger fish like roach.
(d) Pond skaters live on the surface film.
(e) Mosquito larvae are eaten by all fish.
(f) Mosquito larvae swim in the water. When they pupate, they float from the surface film.
(g) Mayfly and stonefly larvae live at the bottom, in mud and on stones.
(h) Larger fish eat mayfly larvae and mayflies that dip as they are flying over the water.
(i) Plankton, water fleas and small midge larvae live in the top layers of water.
(j) Plankton feed on small algae and bacteria that are in the top layers of water. *(20)*

Trace five food chains from your pond web. *(5)*

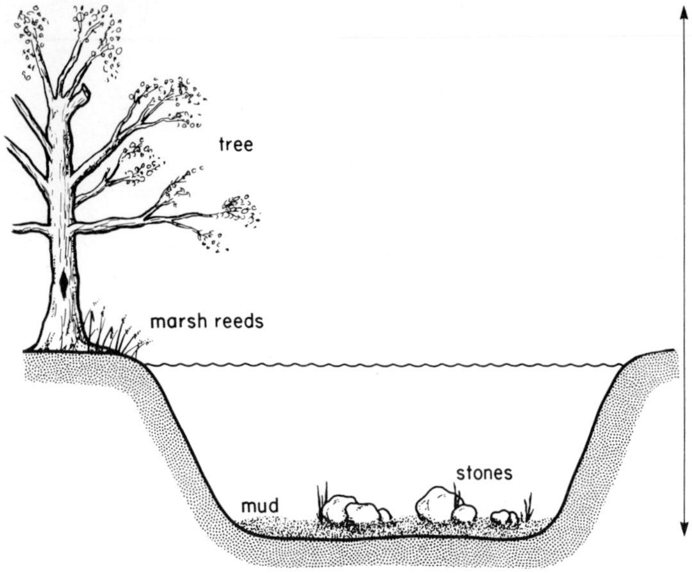

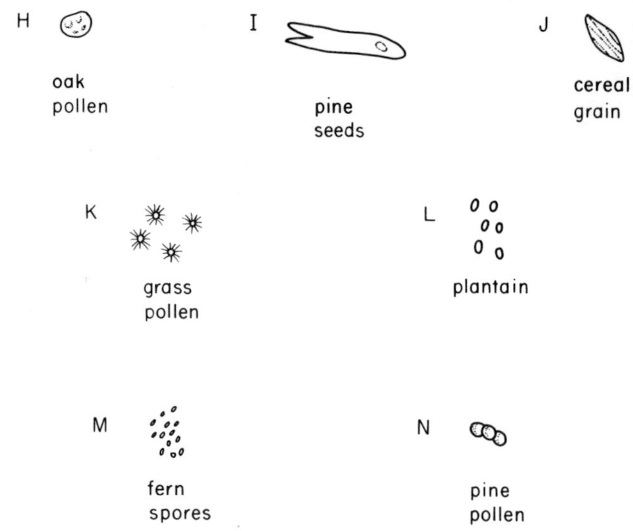

9 Archaeological dating

Some plant remains are very tough and they do not decay away. Seeds and pollen grains are good examples. They are so clear under the electron microscope that they can be identified. If the seeds and pollen in earth samples from archaeological sites are taken for analysis we can find out the kinds of plants that were growing at the time. The age of the sample can be found by C-14 (radioactive carbon) dating.

This is a diagram of some archaeological layers found in an excavation of a village hut circle. Use it to help you to answer the questions.

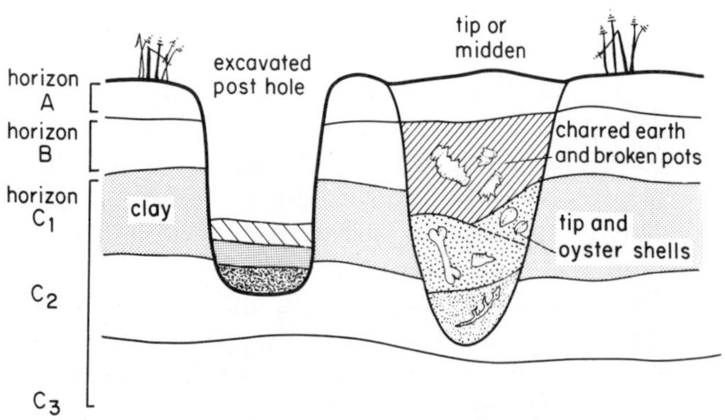

(a) A sample from the bottom of the post-hole contained pollen H and pollen K. At what time of year was the post-hole dug? *(1)*

(b) Could the depth of the midden pit mean that it is older than the post-hole? Say how you would take three samples and use the results to test your answer. *(1)*

(c) Plant material from horizon B was of three types in this ratio: *(4)*

L		K		M
3	:	10	:	1

From this information, describe the nearby landscape at the time. You will have to explain how you worked your answer out. There are several answers. Can we say if one is right?

(d) A cereal seed was found pressed into the clay of a broken pot in the pit. Which layer ought to show other cereal grains and cereal pollen? *(1)*

(e) Seeds of type I were found at the bottom of the post-hole. Could this post have been made of oak? Give two reasons for your answer. *(2)*

(f) Horizon A gave these results from analysis: *(3)*

Pollen H	Pollen N	Pollen K	Seed I	Seed J
20%	10%	30%	2%	38%

Draw a map of the kind of settlement and its surroundings that might have existed at that time.

(g) Describe the changes in this area from the time that horizon C was formed to when it was excavated. *(3)*

(h) The graph below shows how the quantity of C-14 remaining in a sample of material becomes less as time passes. Which layers could be best analysed by C-14 dating? *(2)*

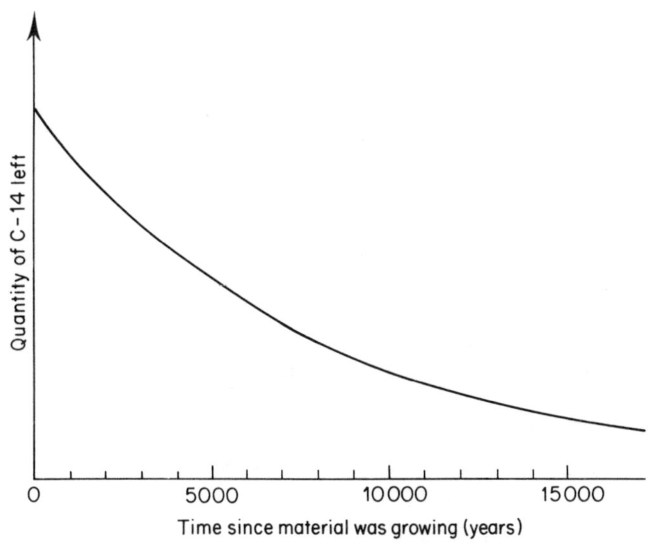

*10 Chemistry of hair

Hair is made of keratin. Keratin is a fibrous protein made of amino acids in a long chain. The chain is coiled in a helix. Hydrogen bonds between amino acids above each other hold the helix in a stable state.

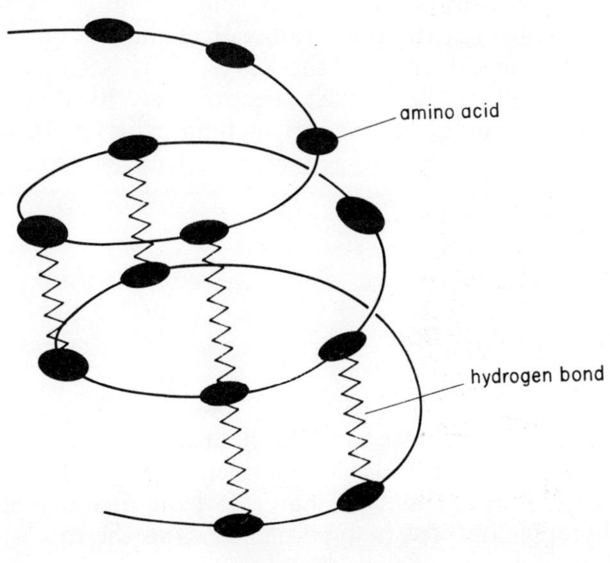

The helical chains of keratin lie next to each other in bundles. These bundles form the hair.

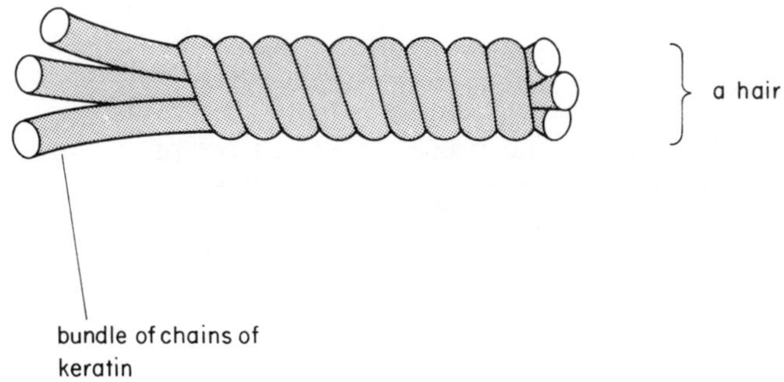

About 12% of the amino acids in keratin are cysteine. This amino acid contains sulphur and has a side branch that can form links with other cysteine amino acids.

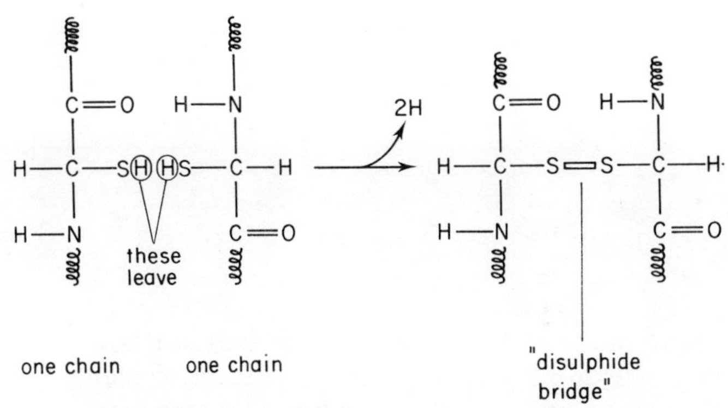

These 'bridges' hold the chains in bundles and give hair its strength.

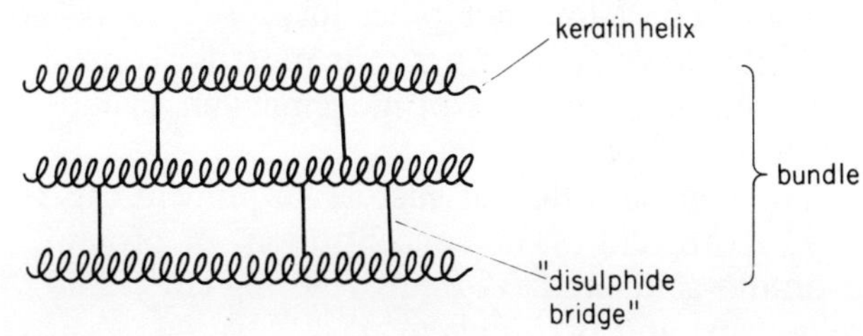

The hydrogen bonds between amino acids of a helix can be broken temporarily by heat and new ones can be formed when the hair cools. If the hair gets damp or is treated roughly then it returns to its original shape. Some chemicals like thioglycollic acid break 'disulphide bridges' enabling hair to be totally reshaped. This takes time. The new bridges that form are permanent and unaffected by heat and dampness. The process is called perming. The perming solution must be washed off and neutralised after it has been on the hair for a sufficient time.

(a) Explain how you think curling tongs might curl hair and why it is not permanent. *(3)*
(b) What might be the main active ingredient of perming solution? *(1)*
(c) Using your knowledge of chemical reactions, what two things could you do to speed up the time needed to perm someone's hair? *(2)*
(d) Why might perming make hair more brittle? *(1)*
(e) The best way of improving the quality of your hair is to eat foods such as eggs, meat or milk. Smearing sulphur compounds onto the hair does not improve its quality. Explain. *(3)*

*11 Sap

Many methods have been used to find out how materials are transported in the phloem of plants. One technique uses C-14 (radioactive carbon) supplied in the form of carbon dioxide. The C-14 is fixed in photosynthesis and the sugars produced are labelled with the C-14 isotope. This can be detected by using a Geiger counter or an X-ray film.

(a) Using one of these methods, describe how you would measure how fast labelled sugars move in a plant stem. (Do not include fixation by photosynthesis.) Draw simple diagrams and use a flow chart for your procedure. *(10)*
(b) Write down four things that must be controlled to compare the rates of movement in two species of plant. *(4)*
(c) The concentration of sucrose in phloem sap is 93.25 mg/cm^3. The concentrations of protein, amino acids and ATP are 1.50 mg/cm^3, 5.00 mg/cm^3 and 0.25 mg/cm^3. *()*

Plant	*Average rate of movement of phloem (cm^3/hr)*
Grass	20
Oak	5
Geranium	25

(i) Which substance forms the greatest part of sap? *(1)*
(ii) Glucose is one of the major products of photosynthesis, but it does not appear in sap. Suggest a reason for this and what might happen to the glucose. *(3)*
(iii) What volume of sap should travel in:
1. grass in 24 hours?
2. oak in 5 hours? *(2)*
(iv) What mass of sap travels in the geranium in 12 hours? (Density of sap 1.035 g/cm^3) *(3)*
(v) The actual amount of sap travelling in grass in 24 hours (1 day) is much less than the theoretical amount. Suggest two reasons. *(2)*
(vi) To compare grass, oak and geranium fairly, what else must you know, apart from the volume of sap and time? *(2)*

(d) The equation for photosynthesis is

$$\text{Carbon dioxide} + \text{water} \xrightarrow{\text{Sunlight}} \text{sugars} + \text{oxygen}$$

H_2[O] - - - - - - - - → (oxygen)

Scientists have done an experiment using an isotope of oxygen, 0–18, to prove that oxygen released in photosynthesis comes from the water taken in at the roots rather than the carbon dioxide taken in by the leaves. This isotope can be detected by a mass spectrometer. Explain what should be done in such an experiment. *(4)*

*12 Lakes

Diagram 1(a) shows how the light intensity at a depth of 5 m in Lake Windermere changes during the year. Diagram 1(b) shows how the quantity of phytoplankton (plant type plankton) in the top 5 metres changes during the year.

Diagram 1(a)

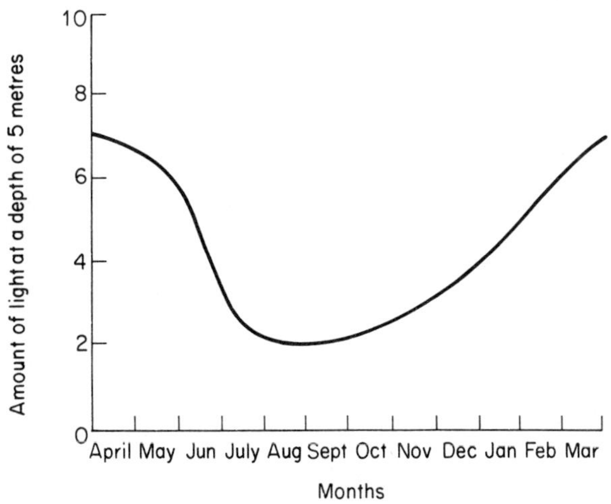

Amount of light at surface = 100 units

Diagram 1(b)

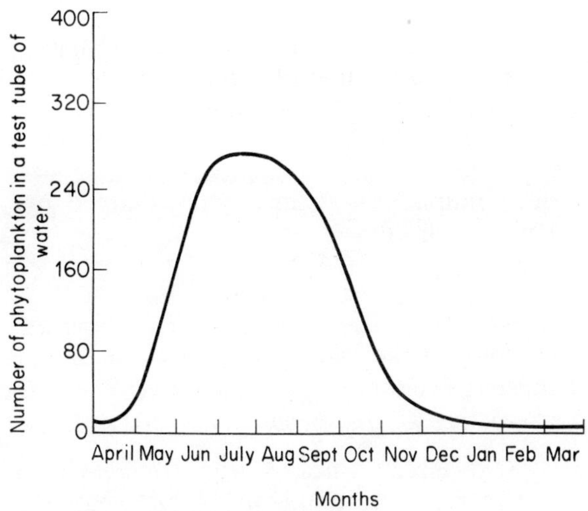

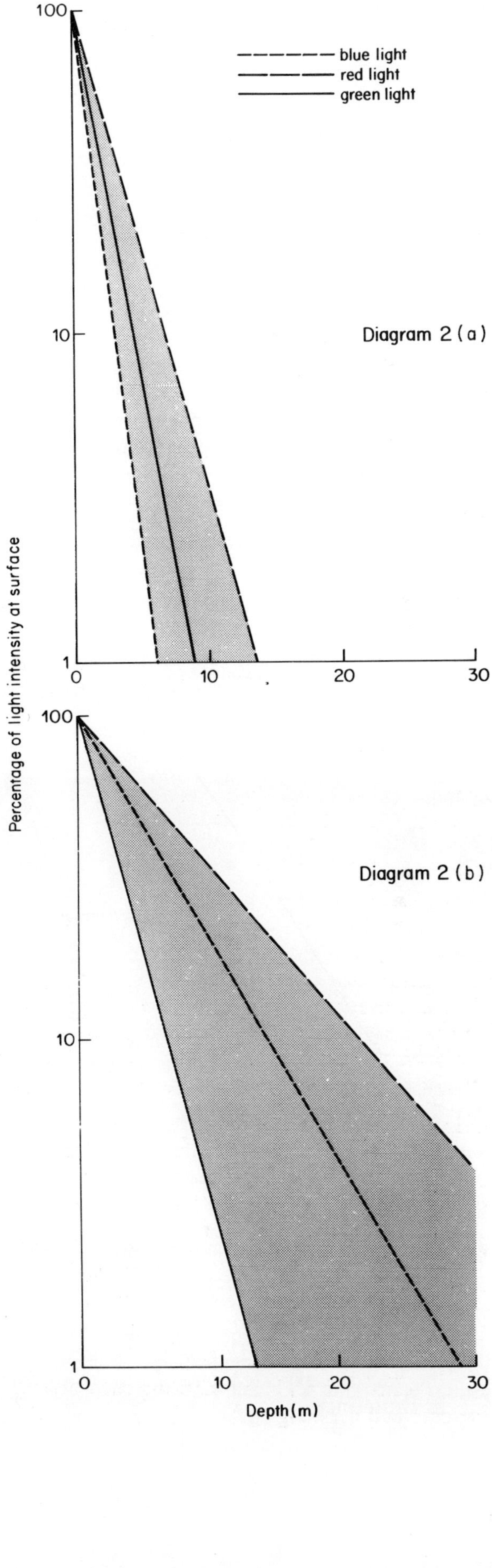

(a) (i) Does the light intensity depend on the quantity of phytoplankton? Explain your answer. *(2)*

(ii) What does the quantity of phytoplankton depend on? *(4)*

(b) Diagram 2(a) shows how light intensity varies with depth in Lake Windermere. Diagram 2(b) shows how it varies in Lake Buttermere.

(i) Which colour of light is absorbed least in these lakes? *(1)*

(ii) Which colour of light do plants require most for photosynthesis? *(1)*

(iii) Which of these two lakes would you expect to have most phytoplankton? Give two reasons for your answer. *(3)*

(c) Freshwater biologists can measure the quantity of nitrogen going into a lake. For Windermere it is thought to be about 500 tonnes per year.

(i) How do you think biologists measure this? *(4)*

(ii) For Windermere, the quantity of nitrogen leaving the lake is about half that entering it. What happens to the rest of it? *(1)*

(iii) Nitrogen can be used up in this way only because the lake has plenty of phosphate. Explain this. *(2)*

(iv) For Buttermere, the quantity of nitrogen leaving the lake is about the same as that entering. Explain how this can be predicted by comparing diagram 2a and diagram 2(b). *(3)*

Diagram 3

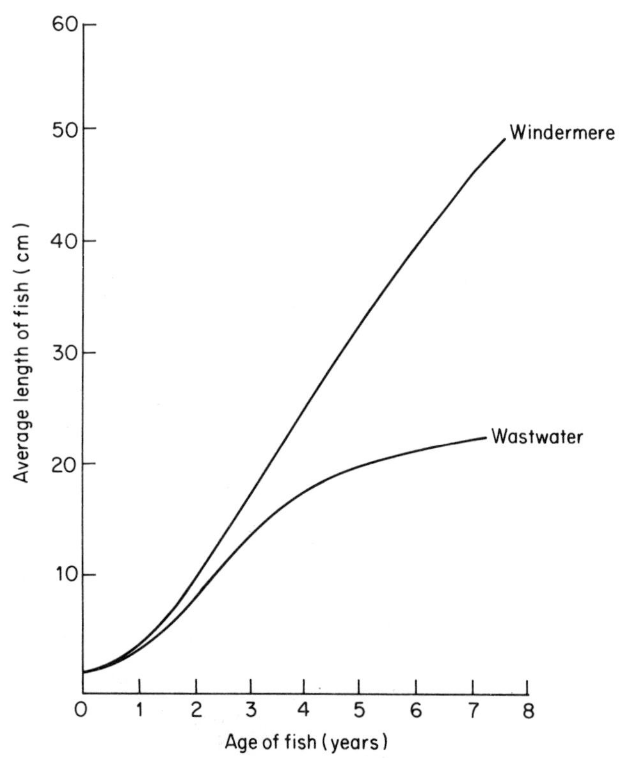

(v) Diagram 3 shows the average growth rate of fish in Windermere and Wastwater. Wastwater has the cleanest water of any of the lakes in the English Lake District. Copy diagram 3 and draw on it a graph showing the expected rate of growth of fish in Buttermere. (2)

(d) Windermere is described as a productive lake. Buttermere is described as an unproductive one. Summarize the characteristics of a productive lake. *(4)*

(e) In 1925 W.H. Pearsall did experiments to find which types of phytoplankton were dominant or sub-dominant in ten of the lakes in the English Lake District. The table below shows some of his results.

		Wastwater	Buttermere	Crummock Water	Haweswater	Bassenthwaite Lake	Ullswater	Windermere	Esthwaite Water
Asterionella	diatoms					d	d	d	d
Tabellaria	diatoms		sd	sd	sd	sd	d	sd	
Dinobryon			d	d	d				
Staurastrum		d							

d = dominant = at times more than 66 per cent of phytoplankton
sd = sub-dominant = at times more than 33 per cent of phytoplankton

Pearsall arranged the lakes in a series. (Wastwater is the most unproductive and Esthwaite Water the most productive.)

(i) At different times of year *Asterionella* and *Dinobryon* are dominant in Derwent Water. Where should this lake be placed in the series? *(1)*

(ii) To what extent would you expect to find *Staurastrum* and *Tabellaria* in Derwent Water? *(2)*

(iii) Wastwater contains only one species of desmid (*Staurastrum jaculiferum*) and no species of diatoms. Ennerdale Water contains seven species of desmid and one species of diatom. Only *Staurastrum longispinum* is dominant and *Genicularia elegans* subdominant. Explain where Ennerdale Water should be placed in Pearsall's series. *(3)*

*13 Freshwater ecology in lakes and lochs

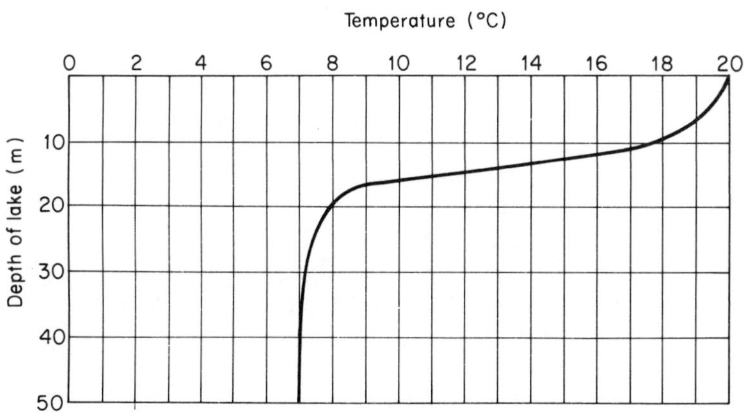

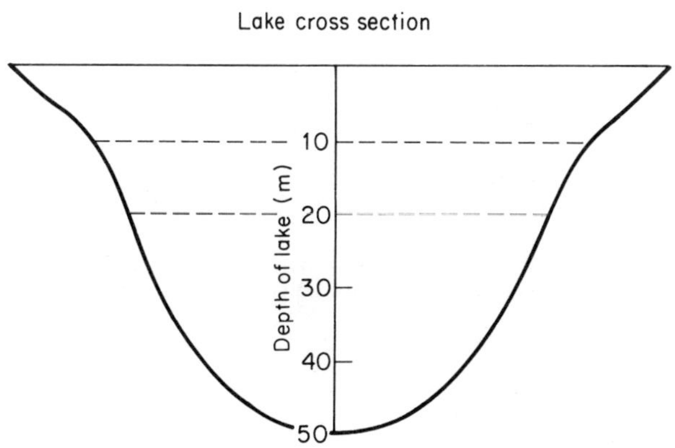

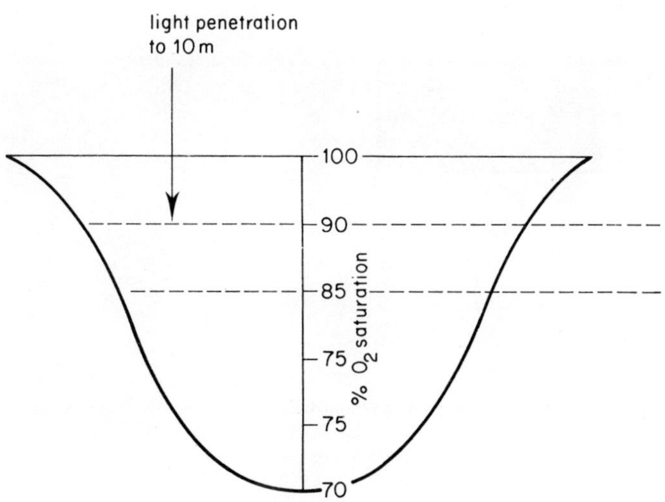

% oxygen (O_2) saturation and light depth penetration

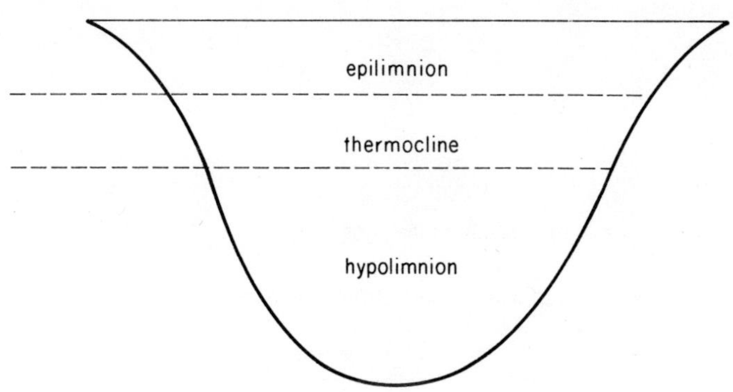

water layers (layers not mixed in summer)

all data recorded in summer months

(a) Between which depths in the lake is the rate of temperature change the greatest?
A. 0 to 10 m.
B. 10 m to 20 m.
C. 20 m to 30 m.
D. 40 m to 50 m. *(1)*

(b) The pH of Loch Lanlish is likely to be
A. pH 3.0.
B. pH 5.0.
C. pH 7.0.
D. pH 8.5. *(1)*

(c) The depth of Loch Calder is
A. 10 m.
B. 20 m.
C. 30 m.
D. 40 m. *(1)*

(d) The depth at which hypolimnion starts is
A. 10 m.
B. 20 m.
C. 30 m.
D. 40 m. *(1)*

(e) The most acidic underlying rock or soil is
A. Marl.
B. Limestone.
C. Granite.
D. Marl overlying granite. *(1)*

(f) Which loch gives the best chance of catching a large trout?
A. Loch Watten.
B. Loch Lanlish.
C. Loch Calder.
D. Loch Skeen. *(1)*

(g) Fish tend to grow better in lochs consisting mainly of
A. Epilimnion.
B. Thermocline.
C. Hypolimnion.
D. No correlation possible. *(1)*

(h) Use your knowledge to explain the oxygen saturation results. (5)

(i) Use the data and your knowledge of biology to explain the average mass of trout found in the different lochs. *(8)*

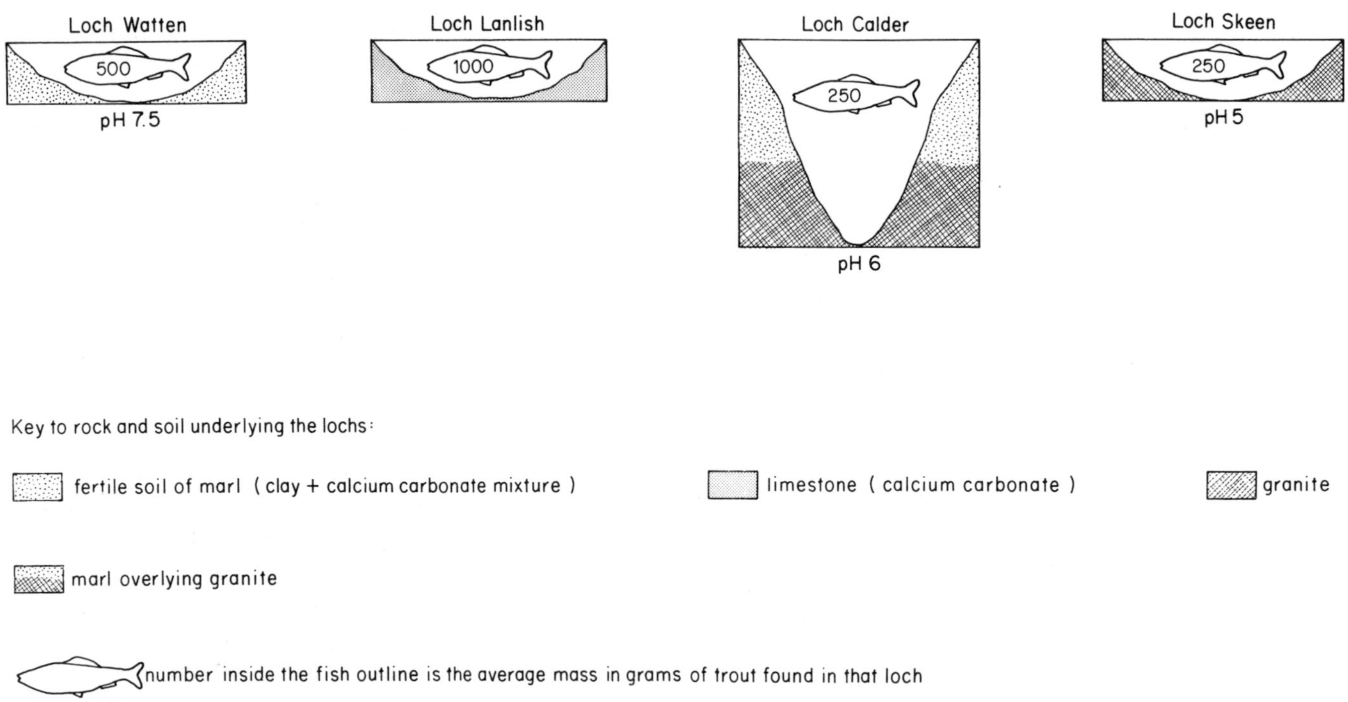

Depths to same scale as before (you need a long ruler)

14

Design a plant which could live and reproduce in the environment described below. Comment on its reproduction and seed dispersal.

Little water available, semi-desert
Nitrogen deficient soil
Sheep and wild goats graze the area from time to time
Insect life is also sharing the environment

Use labelled diagrams and explain your design.

15

Design an animal which could live and reproduce in the environment described below. Comment on its life history.

Extremely hot by day
Water supply is limited, semi-desert
Animal is a night hunter

Use labelled diagrams and explain your design.

*16 Sickle cell trait

Some people's blood contains 'sickle haemoglobin'. Such people are found all over the world and are members of many different races. Sickle haemoglobin is caused by a mutant gene.

Diagram 1

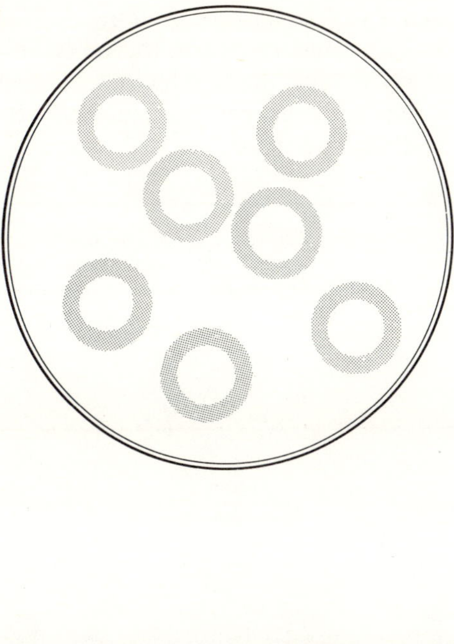

Diagram 2

(a) Diagram 1 shows normal red blood cells. Diagram 2 shows 'sickled' red cells. This is caused by sickle haemoglobin. What differences do you notice in the shape of the cells? *(2)*

(b) Bone marrow and the small blood vessels in muscles allow normal cells to pass through but unusual cells are stopped. What symptoms might this produce in a person with sickled cells? *(5)*

(c) The spleen destroys cells which are an unusual shape. If a person's red cells start to sickle, the person becomes unconscious and may die. Explain why. *(3)*

(d) People with sickle cell trait usually have normal shaped red blood cells. But if there is a shortage of oxygen their 'sickle haemoglobin' crystallizes and their blood cells change shape. During World War 2, often pilots flew aircraft which did not have pressurized cabins. Amongst pilots who crashed their planes, it was found that some had sickled red blood cells. Explain why these pilots crashed their planes. *(4)*

(e) A quarter of children from men and women who both have sickle haemoglobin will have 'sickle cell anaemia' and will die. A half of the children will have 'sickle cell trait'. This means that their red cells become sickle in shape in unusual circumstances only.
Draw a branching diagram to show how sickle cell anaemia and sickle cell trait can be inherited from sickle cell trait parents. *(5)*

(f) In part of Africa 40 per cent of people have sickle cell trait. Look at the maps of Africa overleaf.
   (i) What pattern do you notice between the two maps?
   (ii) The *Anopheles* mosquito carries a parasite called *Plasmodium* which causes malaria. The parasite lives in red blood cells.

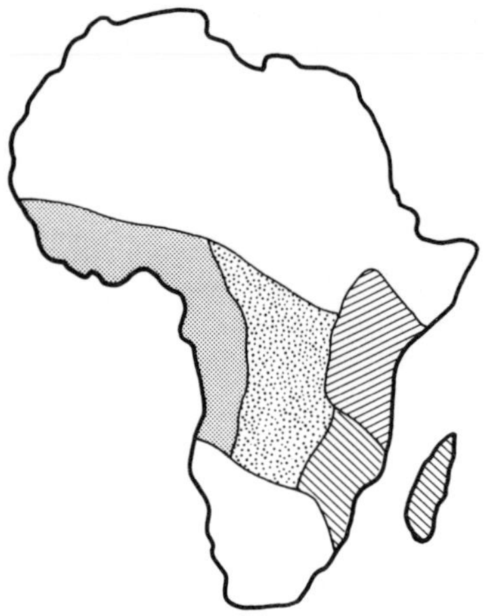

% population carrying sickle cell genes

- more than 25%
- 15–25%
- 10–15%
- <10%

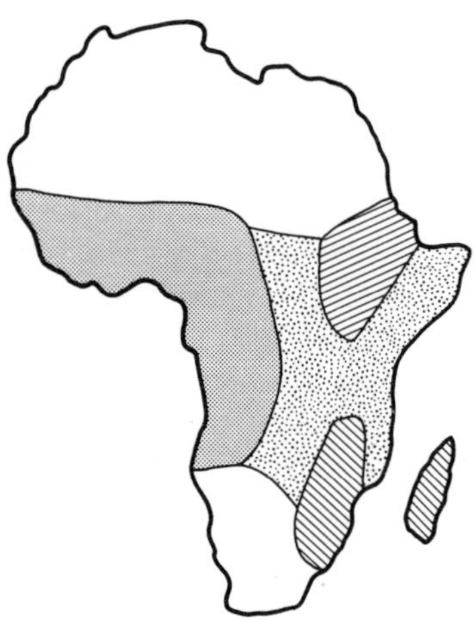

Distribution of *Anopheles* mosquito

- very high
- high
- medium
- low

People often die of malaria, but people with sickle cell trait can get rid of the parasite because the parasite uses up oxygen in affected cells. Explain how sickle cell trait people can get rid of the plasmodium parasite and recover from malaria. *(4)*

(iii) The pattern shown in the maps is due to natural selection. Explain why a high proportion of the people in certain areas have sickle cell trait. *(3)*

(iv) Medical treatment can now control malaria infections. How will this affect the map patterns? *(2)*

*17 Heat pollution

(a) Power stations that use steam turbines depend on the second law of thermodynamics. This states that any process which converts heat energy into mechanical energy must produce waste heat.

The efficiency of a Carnot engine operating between a low temperature T_L°C and a high temperature T_H°C cannot be more than:

$$1 - \frac{T_L + 273}{T_H + 273} \times 100\%$$

(i) Calculate the maximum efficiency of a power station with steam heated to 527°C being condensed to water at 30°C. *(3)*

(ii) Suggest why the real efficiency is actually less than the figure calculated in (i). *(2)*

(iii) Technology has improved the efficiency of steam turbine power stations.

1. Copy and complete the table below to show how the production of waste heat/electrical energy has changed between 1910 and 1970:

Year	*Average efficiency %*	*Waste heat or electrical energy (MW)*
1910	8	11.5
1930	17	
1950	24	
1970	34	

(3)

Diagram 1

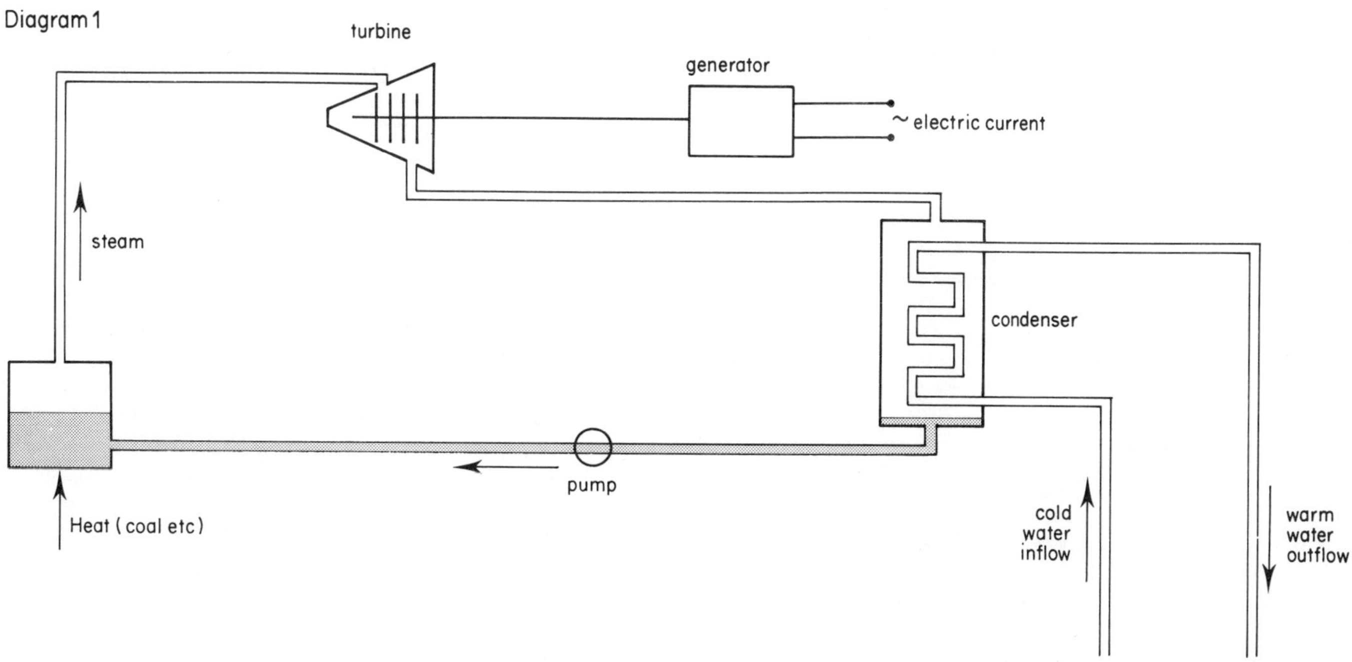

2. How have operating temperatures changed since 1910? *(1)*

(b) A modern power station beside a river (diagram 1) produces electricity at a rate of 800 MW.

(i) If it is 40 per cent efficient, what is the rate at which heat is put into the river? *(2)*

(ii) Use the formula below to calculate the increase in the temperature of the river if it flows at a rate of 120 m^3/s.

$$\text{Temperature rise } ^\circ\text{C} = 0.238 \times \frac{\text{Rate of heat supply (MW)}}{\text{Rate of flow of river } (m^3/s)}$$

(3)

(iii) Your answer to (ii) is an average value for temperature rise in the river. Where will the temperature rise be greatest? *(1)*

(c) Heat pollution is harmful to river life.

Fish A temperature rise of 15°C kills 50 per cent of large-mouth bass in 24 hours, kills stickleback in 35 seconds and salmon in 10 seconds. Increased water temperature increases metabolic rates and the need for food.

Microscopic life

Diatoms grow best at 18–20°C
Green algae grow best at 30–35°C
Blue-green algae grow best at 35–40°C
(Blue-green algae are a poorer source of food for fish than green algae. They are toxic to some fish.)
Bacterial growth depends on temperature. (Bacterial growth affects oxygen supply in the water.)

Physical and chemical properties of water

Diagram 2 shows how the viscosity of water depends on temperature. The rate at which solid particles (including food material) settle is proportional to

$$\frac{1}{\text{viscosity}} \quad \text{(Stokes' Law)}$$

Diagram 2

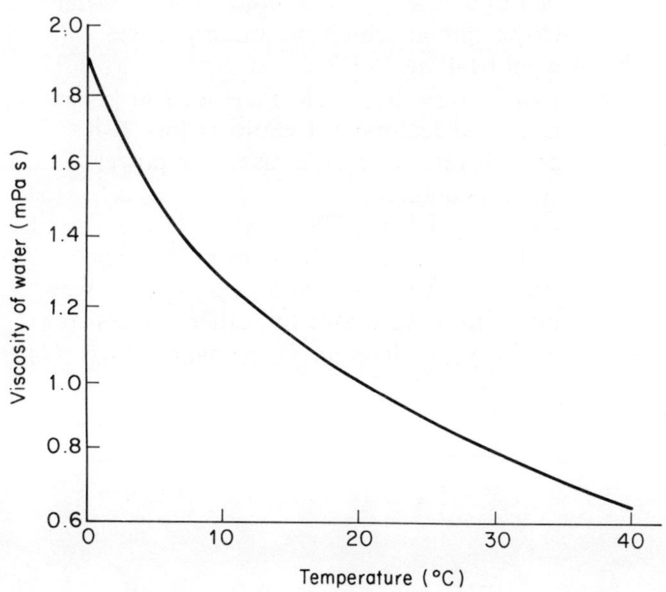

Diagram 3 shows how the solubility of oxygen depends on temperature.

Diagram 3

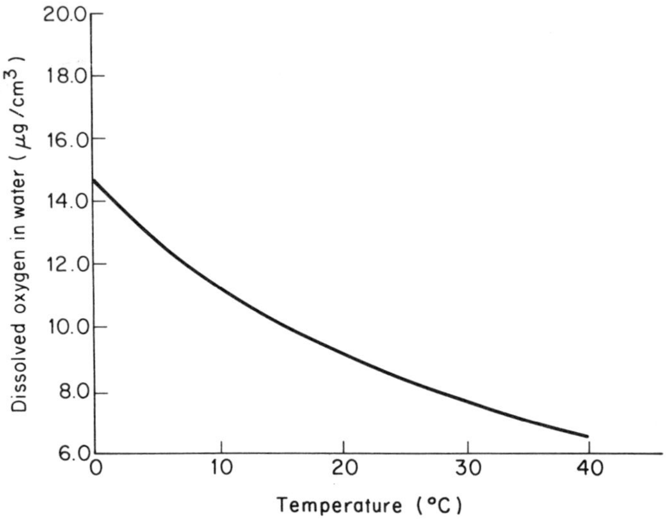

(i) Explain why salmon rivers should not be used for power station cooling. (*1*)

(ii) Suggest a type of fish which is likely to survive in the river described in section (b). (*1*)

(iii) Describe how a temperature rise from 25°C to 30°C is likely to alter the food available to fish in a river. (*3*)

(iv) Give two reasons why there is less oxygen available in warm river water than in cold. (*2*)

(v) A certain food particle falls to the river bed at a rate of 1 mm/hour at 0°C. What is the rate at which the same particle would fall at 25°C? (*2*)

(vi) List factors that could increase fish growth rate, and factors that could reduce fish growth rate, in a river used for power station cooling. (*5*)

(d) (i) On Long Island, USA, oyster farmers used to be able to gather oysters only in the summer. After several power stations were built, they were able to gather oysters most of the year (diagram 4). Explain this. (*1*)

Diagram 4

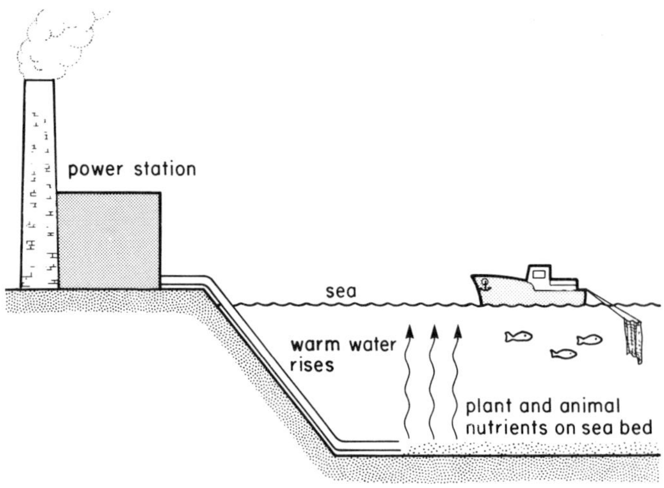

(ii) In some places power stations by the sea have improved fishing. Explain this. (*3*)

(iii) In summer, deep lakes separate into a warm epilimnion and a cooler denser hypolimnion.
Copy diagram 5. Show where cooling water for a power station should be taken from the lake and where it should be put back into it. Explain your answer. (*4*)

Diagram 5

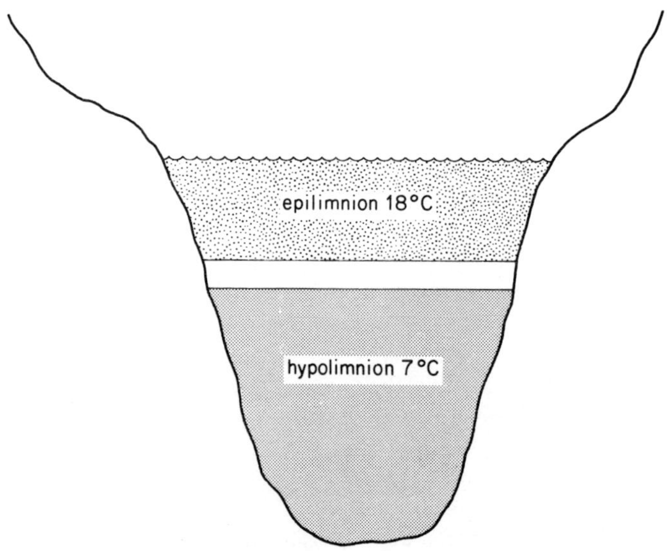

Diagram 6

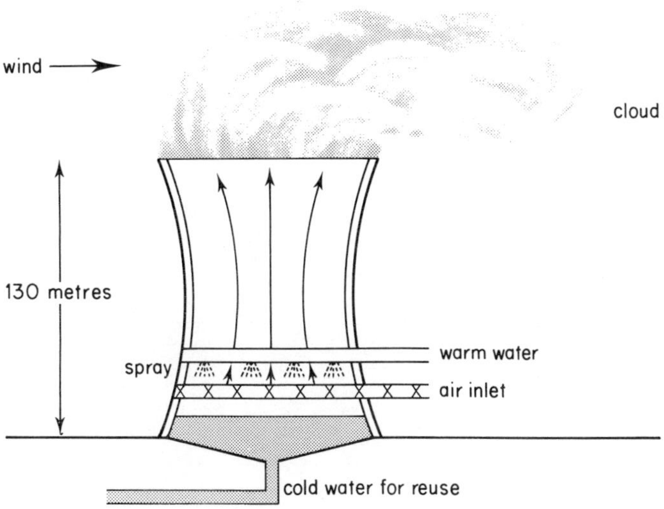

(iv) Cooling towers (diagram 6) cool water by about 10°C by evaporation. About 2 per cent of the water evaporates into the air. One disadvantage of cooling towers is that they make electricity about 2 per cent more expensive. State one other disadvantage and one advantage of using cooling towers. *(3)*

(v) The efficiency of power stations has increased since 1910. But modern power stations cause a greater problem than those in 1910. Why is this? *(2)*

18 Problems with reproductive organs

(a) Read this passage and then answer the questions below.

Some women in their forties and fifties suffer from cancer of the cervix. As part of a routine medical check up, doctors can take a scraping of cells from the cervix (a smear test). These cells are examined under the microscope to see if any are unusual (cancer cells). Most cancers eventually spread to other parts of the body. The cancer cells are carried by the blood and lymph. This is called metastasis. Fortunately it is rare for metastasis to occur with cancer of the cervix. This form of cancer is less common in women who have never had sexual intercourse.

Diagram 1

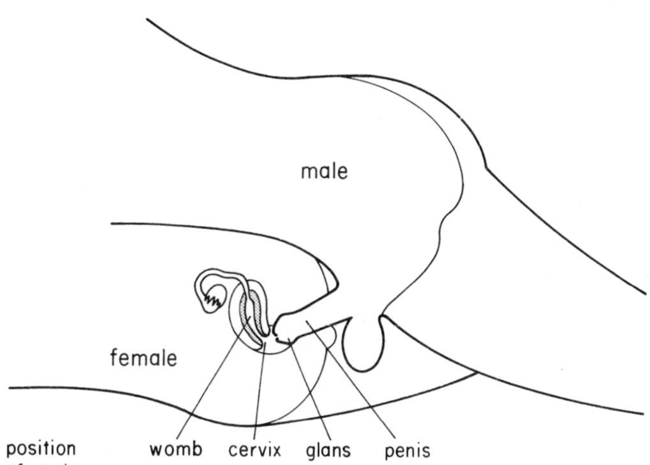

Research has been done on cervical cancer in Malawi. Two tribes occupying the same region do the same jobs. In one tribe it is the custom for boys to have their penis circumcised. (The foreskin which normally covers the glans is cut away.) It was found that married women in this tribe had a lower risk of cervical cancer than the married women in the other tribe. Figures for one area give a risk ratio of 1:4.

(i) Suggest why it is not usual for girls of secondary school age to have a smear test as part of a medical check up. *(2)*

(ii) Using information in the passage, give two reasons why women with cancer of the cervix can often be totally cured. *(2)*

(iii) How does diagram 1 suggest that there could be a connection between cancer of the cervix and whether the partner is circumcised? *(1)*

(iv) Cancer of the scrotum was common amongst chimney sweeps in Victorian times. It has also been associated with certain lubricants used until recently in the engineering industry. Cancers of all types are more common in Cornwall than in other parts of Britain because of radioactive radon gas escaping from the granite rock. In what way did the Malawi research take account of effects not related to male circumcision? *(2)*

(v) Suppose that in a village there are 100 women aged between 40 and 60 with husbands from the circumcising tribe and 100 women of the same age range with husbands from the non-circumcising tribe. It is found that one of the first group has had cervical cancer and four of the second group have had cervical cancer. Is is true to say that the risk ratio is 1:4? Explain your answer carefully. *(3)*

(vi) Technetium is a radioactive element. It is obtained from the fuel rods of nuclear reactors. It can be used to cure cervical cancer.

Diagram 2

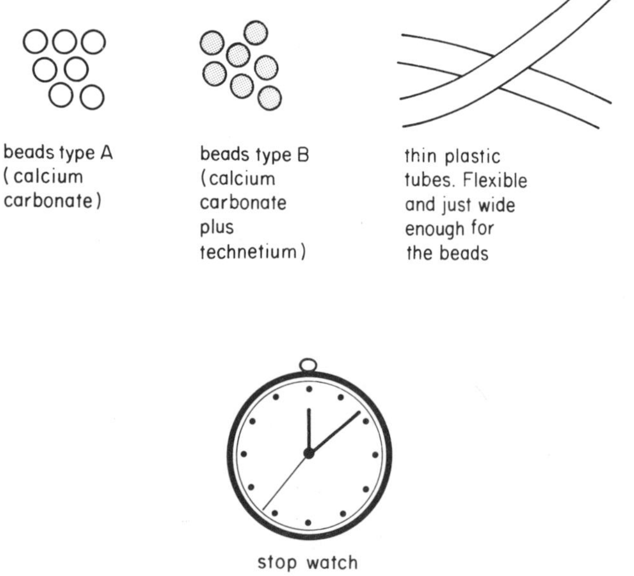

Refer to diagram 2. Describe with the aid of a diagram how this equipment could be used to kill cancer cells in the cervix. *(4)*

(b) Tetanus is a serious disease. It kills over a hundred people each year in Britain. Tetanus causes certain muscles to go into spasms. The muscles operating the jaw are particularly affected so that patients are unable to open their mouths. This is why tetanus is sometimes called lockjaw. Tetanus germs are found in soil. They can be picked up if you cut yourself with a dirty knife. Agricultural workers have a tetanus vaccination regularly. This vaccination contains dead tetanus germs. The body then produces antibodies against these tetanus germs as if they were alive so that if the person has an accident and picks up live germs, the antibodies which have already been produced can destroy them.

If a person who is not immunised in this way is cut with a dirty knife, he or she must have an injection of tetanus serum quickly. A week after the accident it may be too late for the serum to work.

(i) Why does vaccine contain dead rather than live germs? *(1)*

(ii) What do you think serum contains? *(1)*

(iii) Why is the serum unlikely to work a week after getting live germs? *(2)*

(iv) In the Luo tribe in the south west of Kenya, boys have two lower teeth pulled out when they are about fifteen.

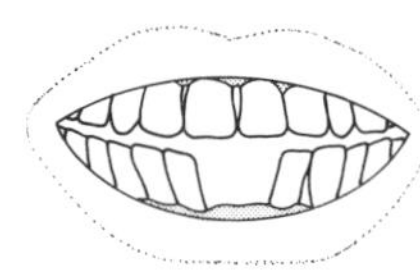

This is an opportunity for them to show that they are brave. They must not make a fuss when the teeth are pulled. It also serves a very useful purpose. Explain this. *(3)*

(c) Keith Masimba grew up in Maragoli. It is on the boundary between the Luo area and the Luhya area. The nearest water hole was about a mile from Keith's home. Luhya boys are circumcised when they are about fifteen. This is a public occasion and it is an opportunity for a boy to show that he can stand pain. A small knife is sharpened on a stone and several boys are cut at the same time.

(i) This practice has many dangers. Which of the following could not be caught by circumcision in this way?
A. Aids.
B. Cancer.
C. Hepatitis.
D. Tetanus. *(1)*

(ii) Keith is a Luhya but he had both Luhya and Luo friends. So he decided to let his teeth be pulled as well as being circumcised. Which should have been done first? Explain your answer. *(3)*

(iii) Occasionally boys die following circumcision. But some doctors in Kenya believe that circumcision helps to keep people healthy in an area like Maragoli. Explain this. *(2)*

(d) Circumcision removes the sensitive foreskin and the glans becomes less sensitive than before. The clitoris is a piece of very sensitive tissue in front of the vulva. It is equivalent to the glans of a man's penis. In central Kenya it was once common for Kikuyu girls to have their clitoris cut off. This is called clitoridectomy or female circumcision. This was an important tribal custom. As with the circumcision of boys, it made people feel that they were grown up.

(i) Some people believe that male and female circumcision reduce the birth rate. Explain whether or not this is a reasonable idea (hypothesis). *(2)*

(ii) How could this hypothesis be tested? *(4)*

*19 Tropical rain forest

(a) Lianes, sometimes called bush ropes, form about 10 per cent of the vegetation in tropical rain forests. Most of them have roots in the ground. They can be more than 70 m in length. They disappear into the lush growth at the tree tops and hang down in loops between trees. They usually produce leaves and flowers only at tree top level. The forest of Kigezi in south-west Uganda is called the Impenetrable Forest. Diagram 1 shows a section through it.

(i) Monkeys can cross a river. Wild pigs cannot cross over. Suggest why. *(2)*

(ii) Tropical rain forests contain many valuable hardwood trees. These are cut down by chain saws and loaded onto lorries. Sometimes it is very difficult to make trees fall once the trunk has been cut through. Suggest why. *(1)*

(iii) It is almost impossible to get into the forest from a road and completely impossible to enter from a river bank because of a dense curtain of lianes. Lianes form more than half the vegetation in these places. Suggest two reasons for the large number of lianes in these areas. *(2)*

(b) Diagram 2 overleaf shows how wet and dry seasons in the tropics depend on latitude. The sun is overhead at the equator in March and September. Ethiopia is north of the equator and the sun is overhead in May and August. Zambia is south of the equator and the sun is overhead in February and November.

(i) It is sometimes said that the rains follow the sun. What evidence is there for this pattern? What other patterns are shown in diagram 2? *(5)*

(ii) It is difficult to do forestry work in wet weather because unsurfaced roads become very muddy. What would be the best time(s) of year to remove timber

1. near Lake Tana in Ethiopia, latitude 13° North?
2. in Kigezi, Uganda, latitude 2° South? *(2)*

Diagram 1

Diagram 2

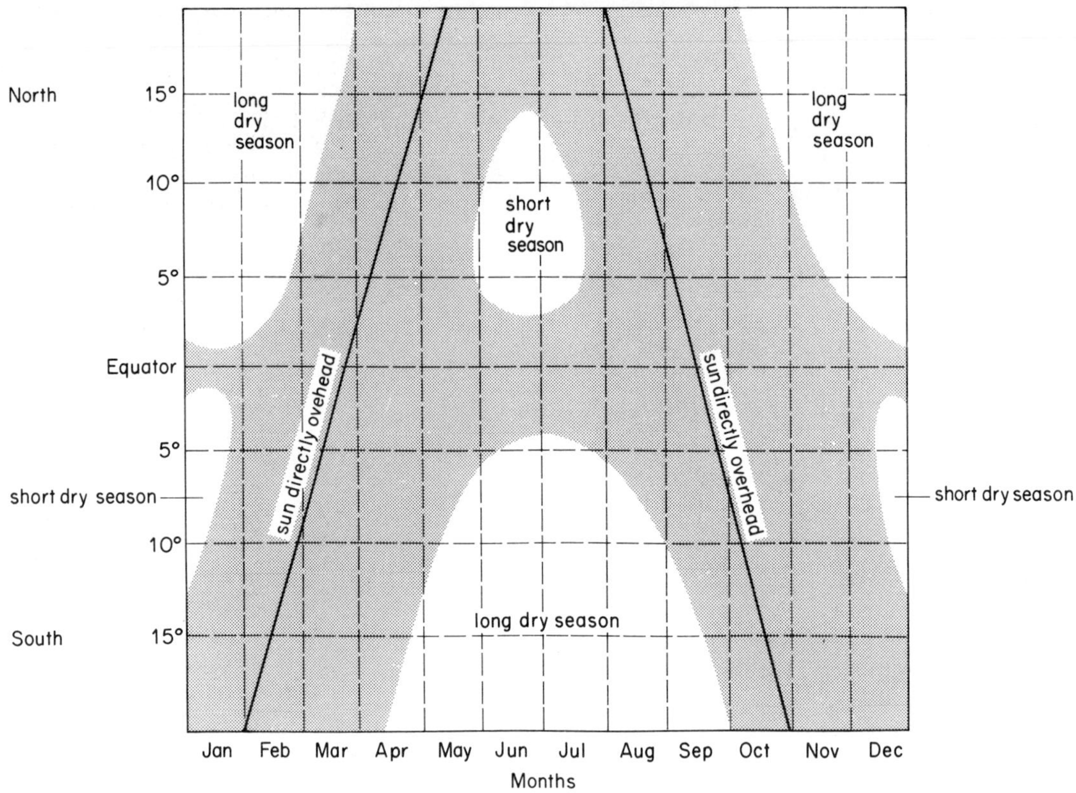

(iii) The rainfall at three places is shown in diagram 3.

Diagram 3(a)

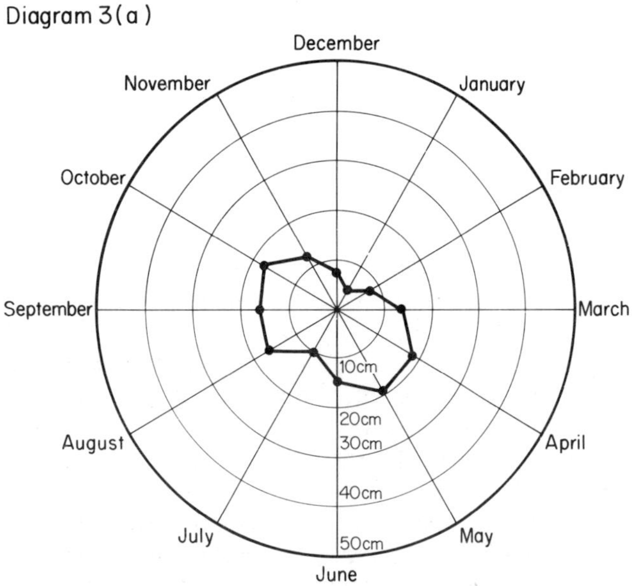

Diagram 3(b)

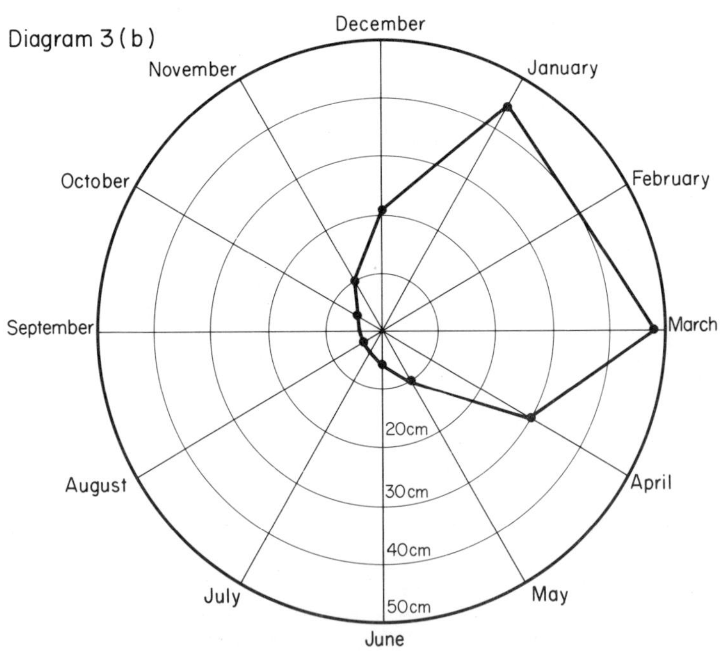

Diagram 3 (c)

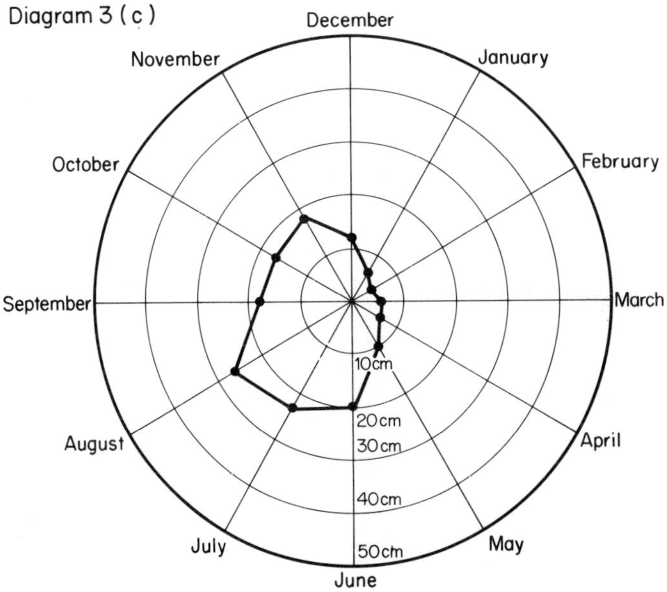

One of these cyclical polygons is for Cairns, Queensland, Australia, latitude 17° South. One is for Port of Spain, Trinidad, West Indies, latitude 11° North. One is for Budongo, Uganda, latitude 1° North. Which is which? Give reasons for your answers. *(6)*

(iv) Tropical rain forests need rain most of the time. They can only exist more than 5° North or South of the equator if they are in mountainous areas. Why is this? *(2)*

Diagram 4

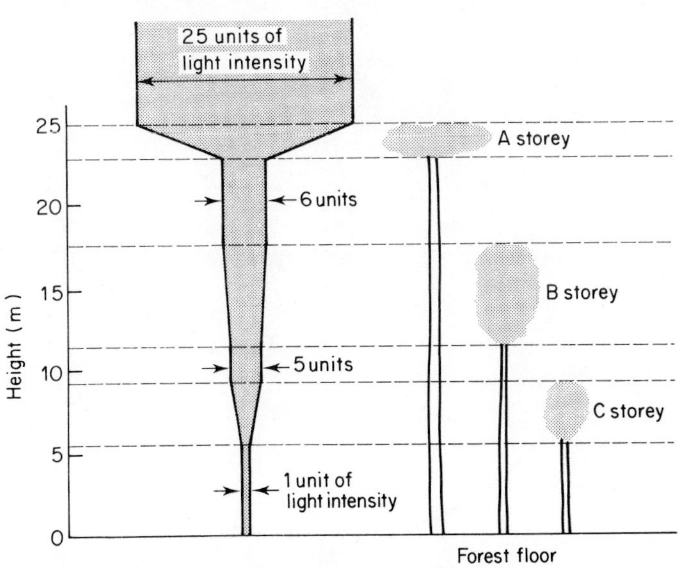

(c) In a tropical rain forest there is vegetation at three levels. Very little grows at ground level. The light intensity has been measured at each level in a Caribbean forest.

(i) Use the information in diagram 4 to draw a cross section of 50 metres of forest showing the amount of foliage at each level. *(7)*

(ii) A forester wants to plant more of the valuable tall growing forest canopy trees in this area. He must allow more light to get to the ground. Would it be more profitable to solve this problem by felling the unwanted B storey or C storey trees? Explain your answer. *(4)*

(d) Diagram 5 shows how the rate of photosynthesis depends on the carbon dioxide concentration for different light intensities.

Diagram 5

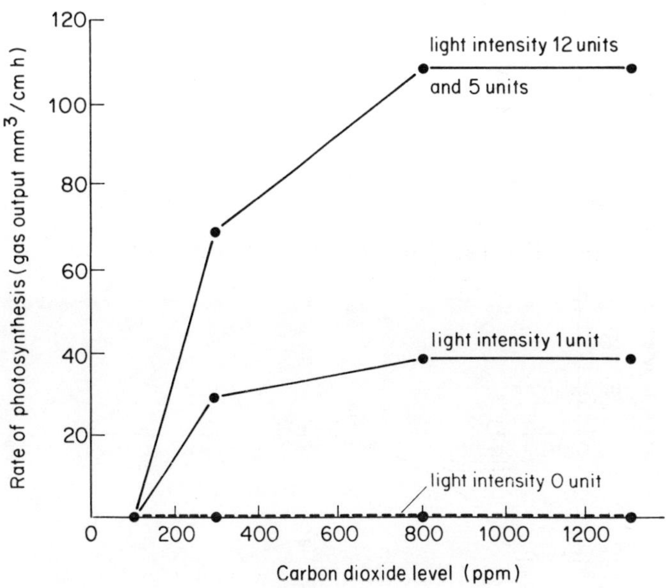

(i) Use the information in diagram 5 to plot a set of graphs with rate of photosynthesis and light intensity as axes. *(4)*

(ii) What levels of light and carbon dioxide are necessary to get the maximum rate of photosynthesis? *(2)*

(iii) The normal concentration of carbon dioxide is 300 parts per million.
In the year 1919, the botanist McLean, working in Brazil, measured the carbon dioxide concentration at the forest floor. He reported 1400 parts per million (ppm) in the evening and 3400 ppm the following morning.

1. Explain why carbon dioxide concentration should be higher in the morning than in the evening. *(2)*
2. Because there is so little light on the forest floor, it is difficult to understand how new plants grow. How does McLean's result help to explain this? *(1)*
3. There is little leaf litter on the forest floor and little humus in the soil. Where has it gone? How does this help to account for the high carbon dioxide levels? *(2)*
4. The carbon dioxide level is about 400 ppm in European forests. Why is it greater in the tropics? *(2)*
5. Diagram 6 is a scatter diagram to show the results obtained by the botanist Evans in Nigeria in 1938.

Diagram 6

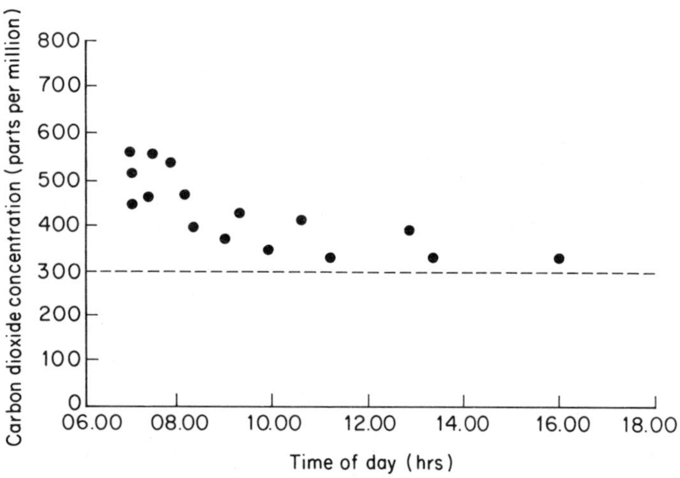

In 1947, Bünning, working in Sumatra, occasionally found values of 2000 ppm very close to the ground.
Compare these results with those of McLean and evaluate the theory that carbon dioxide concentrations are unusually high in tropical rain forests. *(5)*

Talking about the answers

1 Information handling

Information can be given in a table, a graph, a pie diagram, a bar chart or a few paragraphs of writing. Some ways of presenting information are particularly useful in biology. A number of questions in the book use kite diagrams (eg page 2). Pictographs (page 8), cyclic polygons (page 74) and triangular graphs (page 19) are also useful.

You may have to change information from one form to another: from a kite diagram to a map, or from a paragraph of writing to a calculation.

You may have to collect together information from a number of different sources. For this you may have to use your knowledge of biology, to decide what kind of information is relevant, before selecting pieces of data.

Sometimes you have to do arithmetic, use formulas or work out percentages. You have to know what needs to be done and you must be able to do it correctly.

2 (a) What is normal?
Glucose in a person's urine is unusual.
If blood glucose rises above 10 m mol/l there will be glucose in the urine.
So we are looking for the person whose blood glucose level never rises above 10 m mol/l.
This is person N.

Answer **D**

(b) These people are fasted at time 0.
So we want to know K's blood glucose level at time 0.
This is easily read from the vertical axis: 12 m mol/l.

Answer **C**

(c) Following from the explanation in (a), we want to know when M's blood glucose level rises above 10 m mol/l. It is best to put a ruler on the graph. A horizontal line for 10 m mol/l cuts the M graph at 40 minutes.

Answer **B**

3 In a kite diagram, the abundance score is plotted above and below the axis. In direction 1 (going ten metres north), the abundance is 5 for the first 6 metres (21 or more seeds in a quadrat). At a distance of 7 metres, the abundance score is 4 (16 to 20 seeds in a quadrat). No seeds are found beyond 9 metres.

(a) The main cause of this pattern of seed distribution is wind. Wet seeds or seeds covered with ice or snow would tend to drop straight down rather than travel a long way from the tree. But the question asks 'which is the most likely to alter the distribution?' The wind has most influence.
So the data can be interpreted to get information about wind direction.

Answer **D**

(b) At a distance of 10 m, the greatest number of seeds is found north east of the tree (direction 2). So the wind must be coming from the opposite direction: from the southwest.

Answer **A**

(c) We can use data to make 'inferences'. We have already made an inference about the main wind direction. Now we must work out the radius of the tree. When there is no wind, seeds fall straight down. So whatever the radius, some seeds must be found at that distance in all directions.

Diagram (a)

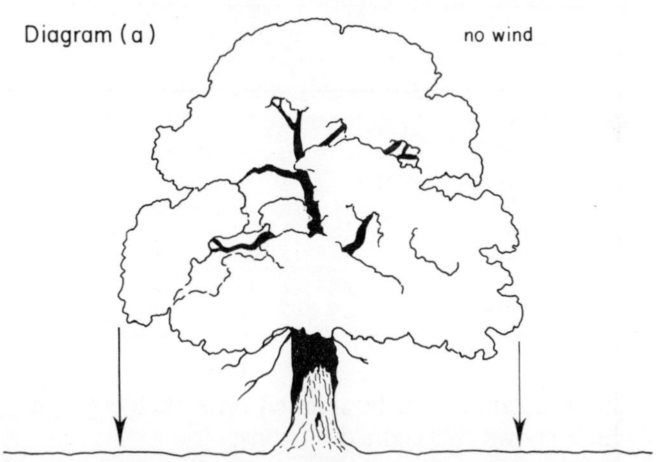

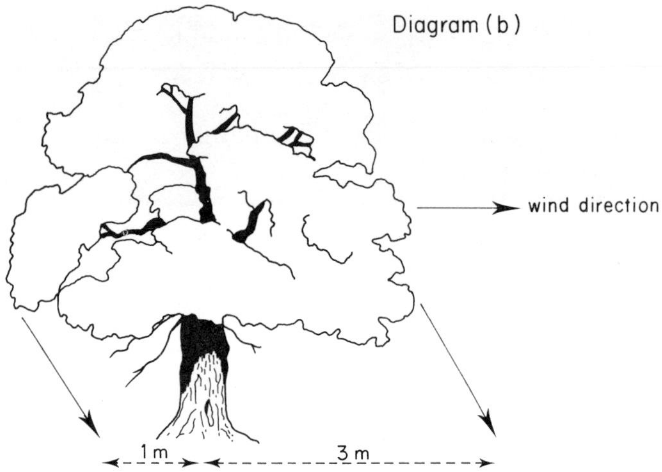

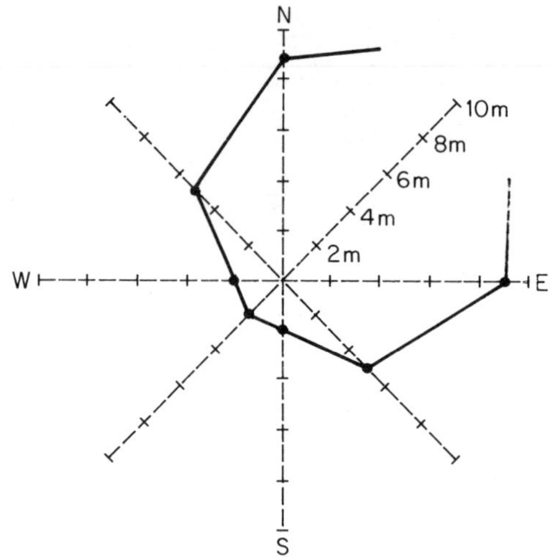

However if the wind blows from all directions, there will be some seeds beyond the radius in all directions. So 2 m would be an over-estimate of the radius.
Statement **D** is simply untrue.
The amount of drifting will depend on the height and length of the branches but there is no evidence that **C** is a correct statement.
A is a correct statement and assumes a situation like that shown in diagram (b) but it does not positively support the conclusion that the radius of the branches is 2 m.
None of the four statements is entirely satisfactory. But the one which backs up the 2 m radius inference best is **B**.

(d) The eight kite diagrams tell us the furthest distance at which seeds are found in each direction.

Direction	*Furthest distance (m)*
1. N	9
2. NE	10 or more
3. E	9
4. SE	5
5. S	2
6. SW	2
7. W	2
8. NW	5

These distances can be marked on a sketch graph which shows the eight directions.

But it should be possible to recognise **A** as the correct pattern without going through all this procedure.

6 Questions **2** and **3** could be answered by using information handling skills alone. In this question, data can be interpreted only if you can match it with facts and theories from your knowledge of human biology.

(a) Energy is produced by respiration in body tissues. Oxygen is carried by red blood cells and combines with glucose to produce carbon dioxide, water and energy. A person who complains of tiredness may have a lack of glucose or a lack of oxygen. If the tiredness continues, there is likely to be a shortage of red blood cells. Such a person would tend to breathe at a faster rate to bring more oxygen to the blood capillaries in the lung. So we have an anaemic patient with a low red blood cell count. This could be V or X. However the person has blood cells which are of normal size. Patient X has cells which are larger than normal (MCV = 125 fl). The person described must be patient V.

(b) Bleeding from the body surface or bleeding below the skin (bruising) can normally be stopped by platelets. These are small sticky cells. If they are exposed to collagen, they clump together to plug a leak in a blood vessel. Mr X and Mr Y both have lower than normal platelet counts. But Mr Y's count is very low indeed. He is the most likely to complain of these symptoms.

(c) This patient's resistance to infection is low. White blood cells fight infection. He must have a low white blood cell count. Mr W fits this description.
(d) Mr Y has about thirty times more white blood cells than normal. He probably has a leukaemia.

18 This passage gives a lot of information. You need to understand biological terms like copulate (have sexual intercourse) and gestation period (time between conceiving and giving birth). You must be able to make connections between different pieces of information in order to work out answers to some of the questions.

(a) The first kittens are born in early April. Gestation is nine weeks. Copulation may be a few days after coming on heat. So mid to late January is the first time. *(1)*
The second time will be nine or nine and a half weeks before some time in July. This is about the beginning of May. *(1)*
(b) The queen makes long, loud and high pitched cries. *(1)*
She rolls on the ground. *(1)*
She creeps along the ground with her tail in the air. *(1)*
(c) The behaviour described in (b) stops. *(1)*
(d) They spray trees, fences etc with urine to mark their territory. *(1)*
They fight other tom cats which come onto the territory. *(1)*
(e) There is no clear answer. It seems to be a basic instinct amongst cats. You must try to suggest a few possible reasons.
Perhaps the mother is now busy feeding a new litter of kittens and cannot be feeding this old kitten. Perhaps the kitten, if female, would compete for toms with the mother.
Perhaps the kitten needs to be taught to look after itself.
(1 mark for each of two reasons)
(f) Two months after some time in July. Mid September? *(1)*
(g) The weather is bad and newly born kittens could die of cold. *(1)*
(h) The cause is the change in the length of the day. But the question asks what *might* control the timing. So any reasonable suggestion is acceptable e.g. temperature or height of the sun. (*Two reasonable suggestions: 1 mark each*)
(i) Whilst they are producing milk their hormone balance suppresses mating behaviour.
They are in poor condition after pregnancy.
They are busy keeping newly born kittens warm and fed and cannot be out calling toms.
These are three reasonable explanations. Others are possible. *(1 + 1)*
(j) July *(2)*
June or August get 1 mark. We are looking for the equivalent of January in England (answer to (a)).
(k) Day length is the same all through the year. (So is temperature). The instinct which prevents cats producing kittens at a cold time of year will not operate. They will come on heat at intervals throughout the year. *(1)*
(l) If the cat was female it could get pregnant at the 'right time' in New Zealand *(1)* and produce kittens at the 'wrong time' in England. *(1)*
(*An answer making any similar point gets two marks.*)
If the cat was male it could jump ship to find a female *(1)* and would have the unpleasant habit of squirting urine on corners unless it was neutered. *(1)*
(*Any two other reasonable points get two marks.*)

2 Problem solving

You may have to start by figuring out exactly what the problem is. Question **4** says that uneaten fish food and fish faeces cause a problem in a lake. Why? Because bacteria increase and use up oxygen, killing other life in the lake.

Sometimes you have to find the cause of a problem. Your knowledge of biology or a study of the data in the question may help you.

You need to understand the reasons why a solution should work. There may be more than one possible solution and you need to be able to select the best from alternative solutions. One may be the quickest, another the cheapest and another the most effective.

A common problem in Biology is designing a fair test. This means knowing what the variables are in a problem situation and controlling them.

There may be many occasions when your problem is an example of a general one. A set of formulas or rules may already exist. All you have to do is interpret the formula correctly for your particular problem.

1 "There will be a fuelwood problem." What is the problem? The population is doubling every twenty-five years. Energy needs are likely to double at the same rate. Unless an alternative fuel is available, the need for fuelwood will double every twenty-five years.

A simple solution might be to plant more land with trees. This is not possible. Only the land between 2500 m and 3000 m above sea level is suitable for growing softwood trees. All the available land is already used for this purpose. This is called a constraint. The solution to the problem must not involve using more land.

There are two wet seasons during which more than three metres of rain may fall. There are two dry seasons during which no rain falls. Softwood trees can be grown on the slopes of Mount Kenya, Mount Elgon and in the Aberdare Mountains. The land is very hilly. Rain washes away both soil and nutrients. None of this information is given in the question. But it is obvious that climate will affect which types of trees grow well. Nutrients are often washed out of soil by heavy rain. Fires present a risk whenever there is dry weather. You will be aware of this background to the problem from your study of biology.

Trees are often planted quite close together, perhaps one metre apart. Not all survive. Eventually they are thinned out so that the distance between them is about two metres. This means removing unwanted young trees. Trees compete for light, water and nutrients. If they are grown too close together, they will grow only very slowly. Many will die. Growing twice as many trees per hectare is not a solution to this problem. (So the answer is **B**.)

Are the other suggestions possible solutions?

There are research stations where different varieties of tree are tested to find out which one is best suited to the climate. Applying the same amount of fertilizer in smaller quantities throughout the year would reduce the effect of nutrients being leached out of the soil by heavy rain. But this takes up the time of forestry workers and is therefore expensive. Preventing forest fires will obviously avoid unnecessary loss of trees. So we can find good reasons to show that these suggestions could help to solve the problem.

3 This is an example of a general problem. Doctors could have to deal with badly burned patients at any time. The solution is given by the formula, the diagram and the table of times.

A slight problem for a doctor is remembering all this. The diagram is sometimes called 'the Rule of Nines'. You will see that the main areas of the body are either 9 per cent or 18 per cent of the total area. The 9s and 18s add up to 99 per cent and the final 1 per cent is the genital area. An area the size of the person's hand is reckoned to be about 1 per cent. Can you think of a system for remembering the table of times?

You must know how to use the general solution to solve a particular problem. Suppose a person has a burnt foot and ankle. The area affected is about a quarter of the leg or about 4.5 hands. This is 4.5 per cent of the body. The mass of the person is 50 kg.

$$\text{Fluid needed per period} = \frac{4.5 \times 50}{2} = 112.5 \text{ cm}^3$$

11 (a) We need to explain how this solution to a problem works.

Pump E draws air through the apparatus. Filter A removes solid material from the air. Some of the solid material could give an acid or alkaline reaction with the indicator. The universal indicator will be a yellow liquid when it is neutral. It will become red as the amount of acid absorbed by it increases. C will receive light which has passed through the indicator B. The colour recorded will indicate how much acid has been absorbed. D measures the amount of air which has been pulled through the apparatus during the experiment. (*One mark for the function of each part.*)

(b) This solution is also used when looking for other problems. It could be used to measure acidity in air due to nitrogen oxides. It could be used to measure pollution by ammonia, an alkaline gas. The indicator would then go to green and then blue.

(c) What variables can affect the colour of the indicator and the intensity of the light recorded?
   (i) The concentration of sulphur dioxide (SO_2) in the air. This is what we need to measure.
   (ii) The volume of air pulled through the apparatus.
   (iii) The quantity of liquid B. A smaller volume would become concentrated with SO_2 sooner.
   (iv) The concentration of indicator in liquid B. If it is more concentrated, less light will pass through.
   (v) The current through the lamp. When a torch battery is almost used up, a dim red light is produced rather than a bright white light.

   Also the filter must continue to remove solid material which could cause incorrect results.
   (Any two unwanted variables controlled: 2 marks)

18 You need to use your knowledge of biology as a background to this problem.

(a) You don't see ladybirds in winter. It is October. This should tell you that ladybirds are about to hibernate. They have solved their problem of finding a place where they will be able to keep warm and protected from frost.
(b) The diagram shows a mixture of marked and unmarked woodlice at C. So not all could have been marked.
(c) A guess which can be tested is sometimes called a hypothesis. It is a problem which can be investigated.
   (i) Perhaps a centipede eats two woodlice per day. We would need to know what proportion of the woodlice were marked. Then we would have to count the number of marked shells produced at F since marking.
   (ii) Perhaps centipedes feast on woodlice then rest. How would you test this?
   (iii) Perhaps clusters of woodlice gain greater protection from centipedes. Could this be tested?

   (Two reasonable and testable hypotheses: two marks)
(d) Can you make some hypotheses about this?
(e) Can you use your knowledge of biology to puzzle this out?
   Larger central gallery made by female beetle to lay eggs. *(1)*
   Smaller lateral galleries made by larvae as they tunnel. *(1)*
(f) These patterns present a problem. Can you use your knowledge of biology to make sense of them?
   (e.g. Large beetle larvae bore between bark and hardwood, eating softer tissues like phloem and bark cambium. Small beetle larvae bore into the hardwood. They may have different enzymes for digesting different plant tissues or different digestive systems.)
(g) Marking would be for the ability to substitute numbers into the formula (2) and to do correct arithmetic (2). This formula is a general solution to problems like this where a sample of organisms is marked. Could you have worked out this formula for yourself?

3 Patterns

You may have to find a pattern amongst the information that is presented. For example, what are the characteristics of members of the bellflower family? In question **2** you must find the pattern from the flower diagrams. You may have to build up information to make a pattern. You do this when you write a tree key. You may have to describe patterns.

Sometimes you need to identify information or behaviour which does not fit a pattern. This can help you to recognise data which are unreliable or which require further investigation.

A pattern can be used to make inferences (drawing conclusions). If a rabbit track covers a deer hoof mark, you can infer that the rabbit crossed the ground after the deer. From information about the pulling force which can be produced by the arm of a dozen boys, you can infer who is likely to do weight training. You can make predictions from patterns. Doctors are often able to make a diagnosis because a patient shows some physiological pattern which is unusual.

5 You need to understand the patterns in the graph before you can use them.

The black bands across the top of the graph show when chemotherapy was given. These need to be linked with the months on the axis which runs along the bottom of the graph. Chemotherapy was given at the beginning of each month for about ten days.

During therapy the number of leukaemia cells falls. The number of normal cells also falls. Between treatment the number of cells of both types increases. Normal cells increase to at least their previous quantity. Leukaemia cells do not increase to their previous quantity. So, during the six months treatment, the level of leukaemia cells gradually falls to zero.

(a) You now have to use the pattern to give information. We are interested in where the solid line for leukaemia cells cuts the dotted horizontal line marked 'remission'. The solid line goes below the remission line for a few days between the times marked 2 months and 3 months.

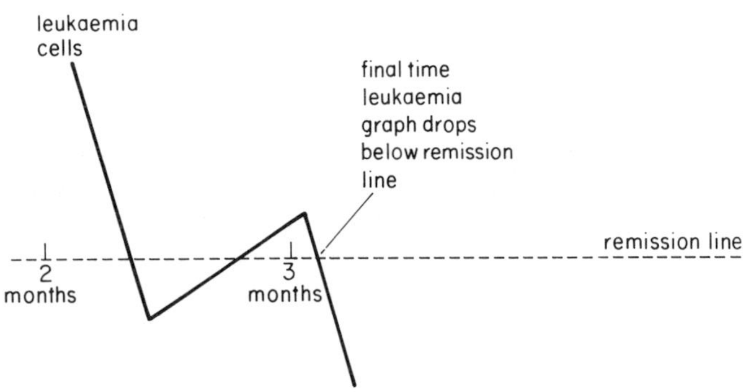

But it is not until after 3 months that the solid line goes below the remission line permanently. Remember the patient is said to be 'in remission' when leukaemia cells are not seen any more. So we want the final time the leukaemia graph drops below the remission line. This happens a few days after 3 months. It is during the fourth month.

Answer **D**

(b) You are being asked to describe the pattern. *During* chemotherapy both normal and leukaemia cells decrease in number.

Answer **C**

But *after* chemotherapy both normal and leukaemia cells increase in number.

(c) You are being asked to 'infer' biological facts from the pattern. Many people incorrectly think that cancer cells divide more quickly than normal cells. This is not true. They divide at a slower rate but do not stop dividing. The control switch to 'off' does not seem to function. Normal cells stop dividing after a certain number of divisions. The graphs show that the normal cells recover at a greater rate than leukaemia cells.

Answer **A**

6 (a) (i) You have to say how this patient's ECG does not fit the normal pattern. The p pattern is the same as in diagram 1 and is normal. The q pattern is present and normal. As the q pattern is normal it is incorrect to say (D) that the qrs pattern is all abnormal. The st pattern is unusual: it is higher than normal. The answer is **B**. Myocardial infarction is present (this means that the heart muscle itself is starved of blood and stops working). This indicates a heart attack.

(ii) Here you are asked to make predictions from the pattern: a diagnosis. The action of the atria is shown by a p pattern when contraction is normal. The p pattern is normal and present. It is not true to say that there is an abnormality in atria and ventricles because atria are normal. Since the st pattern is abnormal and represents ventricular activity, **C** is the best answer. There is not enough information to state which of the two ventricles is abnormal.

(b) (i) You are now obtaining some precise information from a pattern. The r to r interval is 2 large squares. This means one beat every four seconds. Therefore,

No of beats/minute

$$= \frac{60}{0.4} = \frac{600}{4} = 150/\text{minute}$$

The answer is **B**

(ii) Answer **A**

(c) Answer **D**

(d) Which heart is beating fastest? What do we have to look for? Which pattern is most squashed together? Diagram 4 for patient (b).

Answer **B**

10 You need to use a practised routine in order to build up a pattern. We draw arrows in food webs to show the direction in which energy flows. The final diagram usually has the arrows going more or less upwards from producers (plants) to consumers. The highest member of a food web is usually a larger carnivorous animal.
Can you build up the whole pattern?

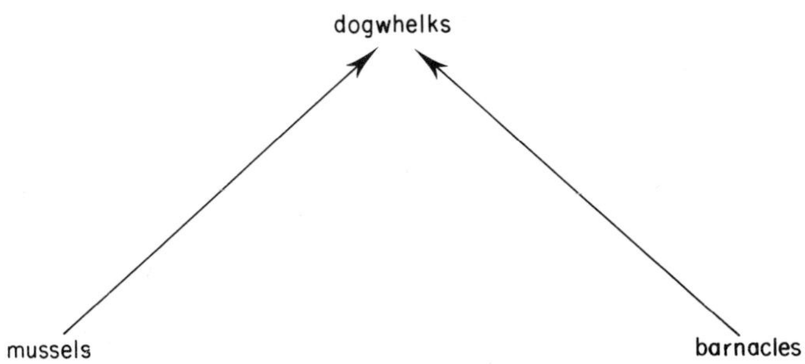

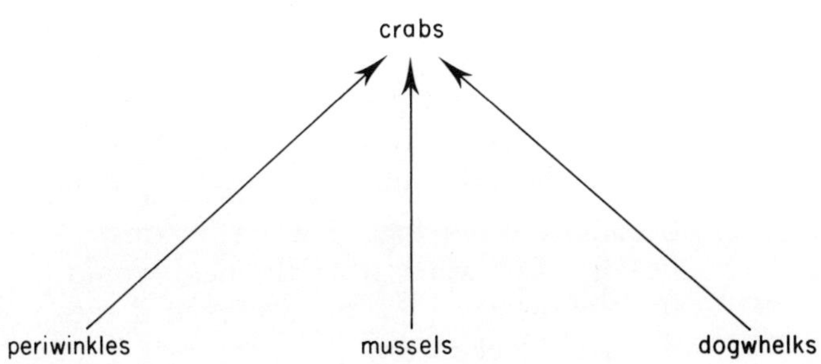

these combine to give

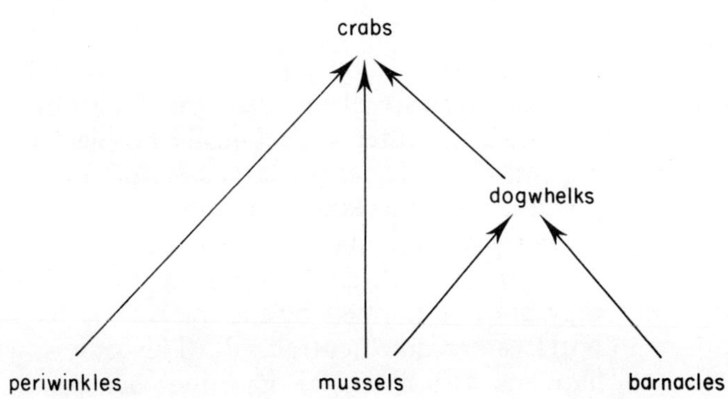

11 (a) This is pattern matching.
Using the key, rabbit and red squirrel are easily recognised. Red deer can be distinguished from sheep by the pairs of dots behind the main print. The final set of prints is for a woodmouse. The dormouse and the coypu are similar but each leaves a tail track. *(4)*

(b) You have to make inferences from this pattern. The acorns are at the left hand side. The animals going towards them are the woodmouse and the red squirrel. *(2)*

(c) This is describing a pattern. The tracks left by the red deer are hoof marks. *(1)*

(d) Again you have to make inferences from the pattern. The rabbit and squirrel prints come in pairs suggesting a series of leaps. The other two animals show 'walking' tracks. *(2)*

(e) A patch of disturbed snow near the acorns could be where an animal sat down. The tracks going to and away from this place are for the woodmouse. *(1)*

(f) Rabbit tracks are marked on top of a deer print. But a deer print has squashed a squirrel print. So the squirrel came before the deer (1) and the deer before the rabbit (1). The woodmouse had time to sit down. But we don't know exactly when it came (1).
So possible answers are:
woodmouse, squirrel, deer, rabbit;
squirrel, woodmouse, deer, rabbit;
squirrel, deer, woodmouse, rabbit;
squirrel, deer, rabbit, woodmouse. *(3)*

4 Evaluation

You will need to make judgements about statements by comparing them with data or with your knowledge of scientific principles. Some facts may have been overemphasised whilst others have been left out.

There is a superstition that you should not eat blackberries after the end of September. Tradition has it that the Devil fell from heaven at this time and landed in a blackberry bush. To mark this event, he has spat on blackberries at the end of September every year and spoiled the fruit. In fact, early frosts spoil the fruit and late fruits are more likely to be infected by raspberry beetle larvae.

There is also a superstition that you should eat rabbits only when there is an 'R' in the name of the month. Is there any scientific basis for this?

In making decisions based on the examination of the evidence, you may feel that some evidence is missing.

You may sometimes have doubts about the reliability of information. Question **13** gives 'information' on the health hazards of bracken which some people may not agree with. We have quoted its source so that readers are able to check it.

Experimental procedures can be criticized. Have all the variables been controlled properly? Could the experiment give fluke results? Will measurements be accurate?

Solutions to problems must be judged. For example (Q2), how safe are alternatives to mothers' milk? Is there any clear evidence (Q**15**) that breast feeding or not breast feeding spoils the shape of the breast?

1 Here we have three solutions to a problem. We must judge the experimental procedure for practicability, (will it work?), and reliability, (is it accurate?), and simplicity, (is it a quick and easy experiment giving a direct result with little calculation?). In order to judge you must be familiar with the apparatus and the size of the quantity being measured. The volume of air in a hen's egg is about 2 cm^3.

(a) The volume of air is given by the fall in water level as the egg is cracked. If it goes from 87.0 cm^3 to 85.3 cm^3, the volume of the air is 1.7 cm^3. (**C**) All the air is likely to be released because it is easy to poke the egg. (**D**)
A difficulty is that hen's eggs are usually wider than a 100 cm^3 measuring cylinder. (**A**)
Another problem is that the scale is usually marked in 1 cm^3 divisions. This makes it impossible to get an accurate result. (**B**)
The two main problems are **A** and **B**.

(b) Burettes are graduated in 0.1 cm^3 divisions. This is possible because they are very narrow. If the area of cross section is 1 cm^2, the 0.1 cm^3 divisions are 1 mm apart. There are measuring cylinders which have 1 cm^3 divisions. But they are very narrow. The measuring cylinder shown in the diagram for plan 2 looks wider than the burette. So it is unlikely that it will be as accurate. As explained in (a), plan 3 is not very accurate at all.

Answer **A**

(c) The egg is broken and all the air must be released. So **D** and **A** are wrong.
No calculations are necessary. But this is also the case with plan 1 and plan 2; so it is not really an advantage.
With plans 1 and 2, rising air bubbles could miss the funnel. With plan 3 the air should escape into the atmosphere. This is the advantage.

Answer **B**

4 (a) Eutrophication is the name given to one form of water pollution. There is excessive growth of algae or other plant material (e.g. water hyacinth). As this decays due to the action of bacteria, oxygen is used up. This increases the biological oxygen debt of the lake. Fish die. It is caused by the presence of too many plant nutrients. Phosphorus, nitrogen, trace elements and carbon dioxide are needed for plant growth. So long as phosphorus is present, for example in lake sediment, excess nitrogen will cause rapid plant growth. Pollution here is probably due to the drainage of water containing nitrates from the fertilized field.

Answer **B**

(b) It is unsatisfactory to seal up the field drains. They are there to prevent the field becoming water logged when it rains heavily. But water from the field would find its own way to the lake even if the drain is blocked up.
The stream supplies water to the lake. It appears to be the only source of water. Diverting the stream would cause the lake to dry up.
It is unreasonable to stop the farmer using fertilizers. They help to produce a greatly increased crop. Alternatives, such as farmyard manure, produce the same pollution problem.
Some kind of filter must be placed in the stream.

Answer **D**

(c) In order to trap all fertilizer run off from the field, the filter should really be placed beyond the field fence. However, of the possible sites marked, **D** is the best.

(d) Nitrates are dissolved in the stream water. They are not present in solid form. So they are not trapped by the straw.
Nitrates are not 'neutralized'. This only happens with chemical opposites: acids and alkalis.

Straw is not a member of the pea and bean family. In any case, nitrogen fixing bacteria convert nitrogen from the air into nitrates and this would not be helpful. Bacteria growing on the straw will consume nitrates as a nutrient.

Answer **C**

(e) The straw probably will silt up. It is more likely to do so in the cage than, for example, tied in bunches which dip into the water along the banks of the stream. Bacteria need oxygen. Oxygen is less likely to be found in deep water.
Corrosion (rusting) uses oxygen rather than producing it. It therefore increases the chemical oxygen debt. The answer is quite simple: the cages stop the straw washing away.

Answer **A**

11 This requires you to evaluate the fitness for purpose of some features of the camel.

(a) (i) Birds have a third, transparent eyelid which protects the eye when the bird is in flight. This is particularly valuable if they have to fly through dirty air. What about the camel? Sand often blows about in the desert. In the Sahara your face is constantly rough with sand sticking to the sweat, and all food is gritty. It is common for trains in north Sudan to be several days late due to sand blown onto the line. Men and women cover their mouth and nose with cloth during a sandstorm. Camels are able to close their nostrils to keep the sand out. *(1)*

(ii) Hair traps air which is a poor conductor of heat. Thick hair acts as an insulating layer. During the day the top of its back will be very hot. It will be at a temperature higher than the camel's body temperature. Thick hair stops this heat from outside being absorbed by the camel's body. *(1)*
At night, thick hair on the camel's back prevents the camel losing heat. *(1)*
During the day the camel needs to lose heat. When it is standing up, it can do this from the belly area because there is little insulation there. *(1)*
(If the camel lies down at night, little heat would be lost from this area.)

(iii) The camel will need to grub for food, usually tough dry grass, in the sand. Their lips need to be tough for this. *(1)*

(b) There are many possible answers to this e.g.

(i) Big feet stop it sinking into the sand. This is clear from the diagram.

(ii) You could find out other things by looking up camels in a reference book.

1. The hump is mostly fat which metabolises to water (and energy and CO_2) during respiration.
2. They do not sweat.
3. Most mammals get rid of nitrogen waste as urea in liquid urine. Birds are easily dehydrated (moving through air causes evaporation and fresh water is scarce). They do not urinate. The white part of bird droppings is a slime which contains uric acid. In the same way, camels economise on water in passing waste nitrogen products.
4. They can go for a week or two without food and can tolerate 30% loss of weight.

16 You have to decide whether the statements made are correct in terms of the information given and your knowledge of scientific principles. Some principles may have been over-emphasized whilst others have been ignored.

(a) Potassium is needed for flower formation *(1)*
and therefore for seed production. *(1)*
Greedy couch grass could have robbed the maize of potassium because it takes up most of its needs by July. *(1)*
Last year the weather was too cold for the maize to put on its growth spurt before June. By then a lot of potassium would have been taken out of the soil by the couch grass. *(1)*

(b) Shepherd's purse, like couch grass, competes with the maize for potassium, light and moisture. *(1)*
Maize does not grow much before June. Shepherd's purse will have gone through two or three life cycles by then, seeding each time. *(1)*

(c) It will not be possible to get rid of shepherd's purse seeds as there will be thousands of seeds. So it must be controlled. *(1)*

Fertilizer added too soon would feed the shepherd's purse. *(1)*
There must be time for the fertilizer to enter the soil *(1)*
or some could 'burn' young plants *(1)*
so correct timing is essential. (*A mark for one correct point in the last sentence.*)

(d) Wild oats do crowd barley. *(1)*
Barley seed is likely to have been collected from a crop which had wild oats in it. Wild oats are seeds which are removed from the plant by the same processes as those used for harvesting barley. So it is likely that wild oats will contaminate the seed. *(1)*

(e) A spray goes everywhere. It would be wise to check that the spray does not harm other plants. *(1)*
There is also the problem of run-off from the field. *(1)*

(f) A biodegradable spray is desirable so that there is no danger of build-up in the soil. *(1)*

(g) Yes, blackgrass can rob wheat if there is low soil potassium. The blackgrass is less affected and will grow better. *(1)*

(h) Camelina has a toxic effect on the leaves and root growth of flax. *(1)*
Once introduced it would continue as a problem by self seeding. *(1)*

(i) Black nightshade seeds germinate after forty years. *(1)*
So after all old plants have been destroyed it is likely that new plants will continue to grow. *(1)*
Control is the answer, probably by human labour.

This advice is useful though some of it needs checking. There is a basis for action. The farmer could order some potassium fertilizer for a start.

5 Synthesis

To tackle real situations in science you usually need to use several skills at the same time. In this section you may need to handle information, solve problems, find patterns and evaluate, all in the same question. You will also have to apply knowledge of biology and cope with situations which look like physics or chemistry.

Multiple choice questions usually test one thing only. So there are no questions in this section which are entirely multiple choice.

Some questions are quite long. A lot of information must be given in order to present a complicated real situation. Some GCSE papers have questions which are designed to take you about twenty minutes. Some of the longer questions in this section will be good practice for GCSE.

4 For this question you will need to remember some facts about the circulation of blood in a mammal.

(a) (i) R is the liver. *(1)*
It is the blood's chemical factory. Digested food from the gut goes to the liver by the hepatic portal vein. Substances dissolved in the blood, e.g. alcohol, can be broken down by the liver.

(ii) Blood is coming from the heart at P. At Q it is going away from the kidneys. You can work this out from arrows on other parts of the diagram. You don't need to remember anything to do this. (2)

(iii) You may remember several relevant facts. For example, the blood flowing out of the kidneys has:
less oxygen
less urea
less water
its salt balance adjusted to normal.
(*A mark for any two correct statements.*) (2)

(b) (i) Blood in state X is pumped to the lungs. It comes from the lungs in state Z. So state X has less oxygen than state Z. *(1)*

(ii) Blood is at a high pressure *(1)*
as it goes from the heart to the lungs. *(1)*

(iii) As it passes by the wall of the gut, blood absorbs amino acids, glucose and water. (Fatty acids are absorbed into lymph vessels.)
(*A mark for any two correct statements.*) (2)

(c) (i) It is called double circulation because there are two paths *(1)*
heart—lung—heart *(1)*
and heart—body—heart. *(1)*

(ii) This is a schematic diagram. You must compare the diagram with the

real situation. Three differences, one mark each. For example, blood does not go down one side of the body and up the other.
The lungs are not in the head.
The 'rest' of the body is all represented by one box, but there are separate blood supplies.
There are no arms and legs in the diagram.
The arms and head are placed with the ribs.
The lungs are not shown as two organs.
The vessels leaving the heart cross over and this is not shown.

(iii) You have to convert the information given in the passage into a labelled diagram. You then have to compare what you have shown with what is now believed correct.
Blood in both ventricles *(2)*
ticked as correct *(2)*
Vapours in pulmonary artery *(1)*
crossed as incorrect *(1)*
Invisible pores *(1)*
between ventricles *(1)*
crossed as incorrect *(1)*
Direction of movement indicated e.g. 'vapours to lungs', 'blood to body' or 'blood through pores'
Any one *(1)*

(iv) The necessary information can be extracted from the passage.
Work with living bodies was forbidden *(1)*
and dead ones did not show circulation. *(1)*

(v) This is correct. *(1)*
His ideas were not accepted in Christian Europe because Islamic and Christian countries did not believe one another's theories at that time. *(1)*

12 (a) You have to interpret the two graphs and make some inferences.

(i) The light intensity at a depth of 5 m must depend on what is in the top 5 m of water. *(1)*
When there is most phytoplankton there is least light. *(1)*
So the light intensity seems to depend on plankton. (However the maximum plankton level is at the end of June and the minimum light is at the end of August. So other things besides plankton must affect light intensity.)

(ii) Phytoplankton is a plant. To grow, it needs light, carbon dioxide and minerals. *(3)*
Graph 1(b) shows that it grows better in the summer. This is because the water is warmer. *(1)*
Also, perhaps, because the light intensity at the surface of the lake is greater.

(b) This is difficult information handling. Data are presented in an unusual way. You will notice that the effect is different for different colours of light. The vertical scale is called a log scale and is not easy to use.

(i) In Windermere, the depth at which intensity is reduced to 10% is about 3 m for blue, about 5 m for red and about 7 m for green. Green is absorbed least. The 'green' line is on the right of the Windermere diagram. It is also on the right of the Buttermere diagram. For both lakes, green light is absorbed least. *(1)*
(This is because phytoplankton, like the leaves of trees, reflect green light.)

(ii) Blue light is most valuable for photosynthesis. *(1)*
Ultraviolet is even better. This is why some plants do not do well in ordinary artificial light inside buildings.

(iii) Light of all colours is absorbed more in Windermere than Buttermere. So the water must be less clear.
Phytoplankton could be the cause. *(1)*
In Windermere, but not in Buttermere, blue light is absorbed most. *(1)*
Phytoplankton would use blue light for photosynthesis. *(1)*

(c) This is problem solving.

(i) The average nitrogen concentration *(1)*
in feeder streams *(1)*
must be measured and multiplied by the volume of water carried by the stream per year. *(1)*
Leaves falling directly onto the lake must also be accounted for. *(1)*

(ii) What can you infer from this fact?
Nitrogen must be used by plants or animal organisms. *(1)*

(iii) You are getting the idea that the nitrogen has something to do with the phytoplankton. Nitrogen is a necessary mineral, so is phosphorus. *(1)* There are usually plenty of phosphates in the sediment of lakes. Shortage of phosphorus would limit phytoplankton growth. If the phytoplankton does not grow, the nitrogen does not get used up. *(1)*

(iv) Little phytoplankton in Buttermere. *(1)* For one of the reasons given before e.g. little absorption of blue light. *(1)* So nitrogen is not used up in the lake. *(1)*

(v) It looks as though fish start to grow at the same rate in both lakes. But the growth rate tails off in Wastwater. There is a suggestion that this has something to do with Wastwater being very 'clean'. We take 'clean' to mean that you can see through the water easily. So not much phytoplankton. If Wastwater is the cleanest, Buttermere must be less clean. Good copy of diagram 3. *(1)*
Buttermere graph drawn somewhere between those for Windermere and Wastwater. *(1)*
It looks as though the fish eat the phytoplankton.

(d) A productive lake has:
phytoplankton in summer *(1)*
so nitrogen gets used up *(1)*
and light intensity is cut down. *(1)*
Fish will develop well. *(1)*

(e) In the early years of this century lakes in Europe were classified as either 'oligotrophic' (not much plant life) or 'eutrophic' (a lot of plant life). W.H. Pearsall introduced the idea of a series linking similar lakes. This idea is now widely accepted.

(i) All lakes in which *Asterionella* is dominant are to the right of Haweswater in the series. All lakes in which *Dinobryon* is dominant are to the left of Bassenthwaite Lake. So Derwent Water must fit between Haweswater and Bassenthwaite Lake. *(1)*

(ii) *Staurastrum* is neither dominant or sub-dominant in any lakes listed to the right of Wastwater, so it is presumably neither dominant nor sub-dominant in Derwent Water.
Tabellaria is dominant in Ullswater but is only sub-dominant in lakes to the right and left of Ullswater and is not even sub-dominant in the lakes at either end of the series. A lake in the position given to Derwent Water should have *Tabellaria* sub-dominant. *(1)*

(iii) Ennerdale Water seems to have more species than Wastwater. *(1)*
It does not contain *Dinobryon* or *Asterionella*. ()
So it must be between Wastwater and Buttermere. *(1)*

Acknowledgements

The authors are grateful to several authors and publishers for permission to use data which has appeared elsewhere. Sources are acknowledged in the text.

Thanks are also due to Julia Bennetts who typed the manuscript of this book and to Caroline Evans for guiding it through to publication. Thanks are due to Jackie Hardie for her thorough comments on the typescript.

The authors and publishers wish to thank Survival Anglia Photo Library for permission to reproduce the photograph on the cover.

Prefixes used in this book

Symbol	Prefix	Meaning	Example	
μ	micro	$\frac{1}{1\ 000\ 000}$	μm	(micrometre)
m	milli	$\frac{1}{1\ 000}$	mm	(millimetre)
c	centi	$\frac{1}{100}$	cm	(centimetre)
d	deci	$\frac{1}{10}$	dm	(decimetre)
k	kilo	1000	km	(kilometre)
M	mega	1 000 000	MJ	(megajoule)

First published in Great Britain 1989

British Library Cataloguing in Publication Data

Science skills: problems in GCSE biology.
1. Biology – Questions & answers – For schools
I. Bennetts, J.
574'.076

ISBN 0 340 49079 9

Typeset in Plantin by Photo·graphics.
Printed and bound in Great Britain for Edward Arnold, the educational, academic and medical publishing division of Hodder and Stoughton Limited, Mill Road, Dunton Green, Sevenoaks, Kent by St Edmundsbury Press, Bury St Edmunds, Suffolk.